W9-AQF-327

BRADFORD
COLLEGE
LIBRARY

GENETHICS
THE CLASH BETWEEN THE NEW GENETICS AND HUMAN VALUES

GENETHICS

THE CLASH BETWEEN THE NEW GENETICS AND HUMAN VALUES

DAVID SUZUKI & PETER KNUDTSON

BRADFORD COLLEGE

JUN 2 1989

LIBRARY

Harvard University Press
Cambridge, Massachusetts
1989

174.25
S968g

Copyright © 1989 by New Data Enterprises Ltd. and Peter Knudtson
All rights reserved
Printed in the United States of America
10 9 8 7 6 5 4 3 2 1

This book is printed on acid-free paper, and its binding materials have been chosen for strength and durability.

Library of Congress Cataloging-in-Publication Data

Suzuki, David T., 1936–
    Genethics : the clash between the new genetics and human values /
by David Suzuki and Peter Knudtson.
       p. cm.
    Bibliography: p.
    Includes index.
    ISBN 0-674-34565-7
    1. Genetic engineering — Moral and ethical aspects. 2. Genetics
— Moral and ethical aspects. I. Knudtson, Peter. II. Title.
QH438.7.S89 1989
174′ .25 — dc 19   88-26013
                                             CIP

Illustrations by Gary Cody

BRADFORD COLLEGE
JUN 2 1989
LIBRARY

OCLC 18442424

To Tara Elizabeth Cullis for her constant support
and faith.

D.S.

To Annette Desmarais, a co-conspirator in my
life, and to my parents, Ruth and Kenneth.

P.K.

TABLE OF CONTENTS

List of Tables *9*

List of Illustrations *10*

Preface *13*

Introduction *17*

Chapter 1: The Legacy of Genes *25*

Chapter 2: Dance of the Genes *48*

Chapter 3: The Dances of the Chromosomes *70*

Chapter 4: The Darwinian Dance: Genes in Populations *94*

Chapter 5: Recombinant DNA: The New Choreography *114*

Chapter 6: Blaming Crime on Chromosomes: The Mystery of the Man with Too Many Ys *141*

Chapter 7: Genetic Screening in the Workplace: Privacy and the Human Genome *160*

Chapter 8: Gene Therapy: The "Moral Difference" between Somatic and Germ Cells *181*

Chapter 9: Biological Weapons: A Dark Side of the New Genetics *208*

Chapter 10: Environmental Damage to DNA: Developing a Sensitivity to the Sufferings of Genes *238*

Chapter 11: Crossing Genetic Boundaries: The Curious Case of the Crown Gall Bacterium *265*

Chapter 12: Maize: In Praise of Genetic Diversity *290*

Chapter 13: Maps and Dreams: Deciphering the Human Genome *316*

Epilogue: Searching for a New Mythology *341*

Glossary *351*

For Further Reading *363*

Index *371*

LIST OF TABLES

Table 1.1: Selected Milestones in the History of Genetics and Eugenics *44*

Table 2.1: The Genetic Code Matching Messenger RNA Codons with Amino Acids *61*

Table 4.1: A Sample Calculation Using Hardy-Weinberg Equilibrium *108*

Table 9.1: Selected Biological Warfare Diseases *210*

Table 9.2 Potentially Hazardous Recombinant DNA Experiments Discussed at Asilomar Conference, 1975 *229*

9

LIST OF FIGURES

Figure 1.1: Evolutionary Clock *27*

Figure 1.2: Mendelian Cross Showing Dominance *37*

Figure 1.3: DNA Double Helix *43*

Figure 2.1: A Human Somatic Cell *49*

Figure 2.2: The Structure of DNA *51*

Figure 2.3: Flow Diagram of Central Dogma *54*

Figure 2.4: Protein Synthesis *56*

Figure 2.5: DNA Replication *66*

Figure 3.1: Mitosis in a Simplified Cell Having Only Two Pairs of Chromosomes *73*

Figure 3.2: Human Chromosomes *75*

Figure 3.3: 24-Hour Clock Showing Stages in the Life Cycle of a Cell *81*

Figure 3.4: Oogenesis and Spermatogenesis *84*

Figure 3.5: The Stages of Meiosis *85*

Figure 3.6: Meiotic Crossing Over of Chromosomes during Prophase I *87*

Figure 4.1: Two Forms of Peppered Moth on a Soot-covered Tree *100*

Figure 5.1: An Overview of the Gene Cloning Process   *117*

Figure 5.2: The Restriction Enzyme Activity of *Eco*RI   *120*

Figure 5.3: Steps in Creating a Human Gene Library   *133*

Figure 5.4: DNA Sequencing Technique   *135*

Figure 6.1: Meiotic Nondisjunction Leading to XYY
Male   *143*

Figure 6.2: Karyotypes of XYY and Turner's Syndromes   *146*

Figure 7.1: Steps in Amniocentesis   *167*

Figure 8.1: Microinjection of Rat Gene into Fertilized
Mouse Egg   *182*

Figure 8.2: Normal and Sickled Red Blood Cells   *187*

Figure 8.3: Normal and *S* Hemoglobin Molecules   *190*

Figure 8.4: Basic Steps in Gene Therapy for Sickle-Cell
Anemia   *195*

Figure 9.1: Steps in Cloning a Human Vaccine   *226*

Figure 10.1: UV Damage and Repair Mechanism   *243*

Figure 10.2: Excision-Repair System   *245*

Figure 10.3: DNA Damage from Ionizing Radiation   *253*

Figure 11.1: Life Cycle of Crown Gall Bacterium   *266*

Figure 11.2: Evolutionary Tree Showing Prokaryotes Emerging before Eukaryotes   *270*

Figure 11.3: Insertion of Recombinant Genes into Tobacco
Plant Using Crown Gall Bacterium Vector   *276*

Figure 12.1: Cobs of Six Types of Maize   *292*

Figure 12.2: Mayan Maize-Bearing God with a Headdress of Cobs *298*

Figure 12.3: Life Cycle of Maize *305*

Figure 12.4: Maize Double-Hybrid Cross *308*

Figure 13.1: Human Pedigree Showing Inheritance of Red-Green Color Blindness *321*

Figure 13.2: Somatic Cell Hybridization Technique *324*

Figure 13.3: Human-Chimpanzee DNA Hybridization Experiment *333*

PREFACE

Genethics is an exploration of the clash between modern genetics and human values. Designed to be accessible to nonscientists, it is both an introduction to the underlying biological principles of the new genetics and a search for unifying ethical themes that can help individuals navigate through the uncharted, often treacherous waters of genetics and morals. The title itself is a novel, "recombinant" word that splices the words "genetics" and "ethics" together to capture their conceptual inseparability.

Despite the undeniable complexities of the ethical issues surrounding modern genetics, we believe that it is possible to offer lay readers something more than another scholarly treatise of agonizing pros and cons or the highly technical musings of scientists. Instead of presenting one more account of the seemingly irreconcilable moral dilemmas arising from genetics, we propose a number of concrete moral "solutions."

Genethics represents a search for broad, lasting moral guidelines — gleaned from complex, real-life ethical issues in genetics — that are at once imaginative, humane and scientifically sound. These moral guidelines, or genethic principles, as we call them, are not intended as final answers to every problem. On the contrary, we offer them as provisional answers that can at least serve as a scaffolding for more meaningful and precise answers that may emerge during the decades of scientific discovery that lie ahead.

For this reason, we make no pretense that our moral arguments are completely objective. The genethic principles we present embody a distinctly humanistic point of view — one that we believe is shared by many thoughtful scientists and nonscientists. *Genethics* differs dramatically from other books on the topic by replacing abstract philosophical debate of moral pros and cons with scenarios or case studies designed to serve as scientific parables, each selected to reveal the single genethic principle that is the organizing principle of each chapter.

In our fragmented society, the power to apply new scientific knowledge tends to reside not in the individual but in vast corporate and military organizations that possess the resources and expertise to harness novel techniques. In our view, motives of profit and military power should not be allowed to shape our society's technological priorities. By providing a foundation of scientific understanding, *Genethics* sets out to empower ordinary people — many of whom understandably tend to shy away from science — to make their own ethical choices in genetics. Everyone must share responsibility for the decisions that will increasingly shape the genetic future of our planet.

Molecular genetics is by nature an incredibly complex subject. And we realize that — despite our best efforts to render some of its subtleties understandable to the vast majority of people who have had little or no university-level training in the life sciences — our decision to refrain from condescending or cartoonish descriptions of intricate hereditary process might make portions of *Genethics* difficult for some. To minimize frustrations, it is important that readers grasp the basic organizational plan of the book.

Genethics includes an introduction and 14 chapters. The Introduction presents a personalized vision of the limits of science and the proper role of scientists in modern society. The first five chapters of the book are intended to serve as a short course, or primer, on selected principles of modern genetics that underlie the discussions of ethical issues in the remaining chapters.

While we strongly encourage a careful reading of these preliminary chapters, we suggest that readers who find themselves bogged down in this section should not hesitate to pause, place a marker at that page and proceed directly to the less technically demanding ethical discussions beginning with chapter 6.

Chapter 1 offers a panoramic view of the history of genes — their primal role in the origin and evolution of all life on earth and in human attempts to make sense of often invisible hereditary mechanisms.

Next, in chapters 2, 3 and 4, we focus on the biological processes by which genes carry out their functions at the molecular, cellular and population levels. We view the activities of genes through the metaphor of dance — as elegant, exquisitely ordered, almost ritualized patterns of movement in space and time.

At the highest level of magnification — beyond that of the most powerful electron microscope — the gene's dance might be called a molecular dance, as genes orchestrate the orderly synthesis of molecules essential to the life processes of the cell and transmit replicas of themselves to the next generation of cells. At the next lower level of magnification, a gene's dance could be called chromosomal, in reference to the ritualized, visually symmetrical movements during cell division of the gene-bearing bodies known as chromosomes. Finally, from a more distant perspective, a gene might be seen to dance to an evolutionary beat — over geologic time and across great geographic distances in populations.

The last primer chapter, chapter 5, describes selected techniques in genetic engineering that have only recently allowed our species to "choreograph" the dances of individual genes to satisfy our own real or perceived needs.

Each of the eight chapters making up the second part of *Genethics* is designed to serve as a genethic parable — an independent search for a unifying moral imperative within a single crucial area of human responsibility for genes. Together, they represent the first fruits of our search for humane, broadly applicable ethical themes in genetics concerning issues such as genetic screening, gene therapy, biological weapons, environmental damage to DNA, crossing evolutionary boundaries, genetic diversity and genetic maps. While most readers will read these chapters in sequence, they could, in principle, be read in any order.

Finally, the Epilogue offers a brief review of the moral themes presented in *Genethics* and offers suggestions for expanding our search for new, more meaningful genethic principles to guide us in the years ahead. Throughout the book, scientific and technical terms appear in italics on first mention and are defined in the glossary at the end of the book. A list of suggested references for each chapter is also included at the end of the book.

We express gratitude to those scientists who took time from busy schedules to engage in freewheeling preliminary discussions of the themes presented in *Genethics*. They include: George Wald, Harvard University; Stephen J. Gould, Harvard University; Jonathan Beckwith, Harvard Medical School; and David Baillie, Simon Fraser University. Special thanks go to Jan Kraepelien and Annette Desmarais for their support and for thoughtful conversations on this topic extending back many years.

The following reviewers generously offered their expertise during the revision process: Jonathan Beckwith, Harvard University; Milton Gordon, University of Washington; Harold Kasinsky, University of British Columbia; Jack Keene, Duke University; Richard Palmiter, University of Washington; and R. C. von Borstel, University of Alberta. While the contributions of these and others to *Genethics* were invaluable, we take full responsibility for the accuracy, content and viewpoints expressed in this book.

Finally, we gratefully acknowledge the efforts of our excellent editor, Nancy Flight, for her painstaking attention to detail without sacrificing the spirit and occasional stylistic leaps of this book. Thanks also to the staff at Stoddart Publishing Company for their steadfast support for this undertaking. We also express our thanks to the staff of the Woodward Library at the University of British Columbia and at the Hastings Institute of Society, Ethics, and Life Sciences, Hastings-on-Hudson, New York, for their assistance. A portion of the research and writing of *Genethics* was funded by Government of Canada: Science Culture Canada Program, and the Ontario Arts Council.

INTRODUCTION

Scientific Explosion

Newspapers and magazines regularly announce the latest developments in science and technology — surrogate mothers, the greenhouse effect, the strategic defense initiative, gene transplants, PCBs, dioxins, artificial intelligence, extraterrestrial intelligence. The familiarity of these arcane terms reflects the extraordinary power of science and technology in shaping our lives and society today. This power is a very recent phenomenon. In fact, from invention to widespread application, the entire history of such revolutionary technologies as automobiles, airplanes, antibiotics, telecommunications, nuclear energy, space travel, computers and oral contraception is encompassed within the span of a single human life.

The application of scientific discoveries and inventions has brought spectacular improvements in our health, material wealth and comfort in a remarkably short period of time. Yet immense global problems such as pollution, overpopulation, the proliferation of nuclear arms, the extinction of numerous species, desertification, deforestation and the destruction of natural habitats can also be traced directly to many of those same technologies. This dual legacy of immediate, powerful benefits on the one hand and unexpected deleterious consequences in the long term on the other hand must be recognized by all members of society today.

17

Often, unanticipated consequences in the long run outweigh the short-term benefits.

Nowhere is the need for the understanding of the general populace more urgent than in the young field within biology called genetics. Genetics is the study of inheritance. In less than a century, geneticists have defined the laws governing heredity and located the hereditary blueprint in a type of molecule within chromosomes. By deciphering the chemical code in which these units of inheritance, called genes, are written, scientists have acquired powerful tools for manipulating the messages that dictate our physical makeup. For the first time in history, it is within our power to design life by deliberate human intervention. The possibilities and implications are awesome, but we lack historical, social and cultural guidelines that can lead us through this uncharted territory.

Fragmentation of Knowledge
In the past, people lived within the comfortable confines of a worldview in which everything belonged and fitted together to "make sense." Part myth, part accumulated experience, part insight, part superstition, a worldview is a holistic construct within which all is interconnected. Thus, for example, the birth of a two-headed calf was once interpreted as a portent of an extraordinary event to come or as punishment for something in the past, because nothing occurred in isolation. That view has changed radically. Today the print and electronic media assault us with a vast amount of information presented in snippets to fit an appropriate space or time slot. This proliferation of information and a parallel increase in areas of expertise, each with its own specialized jargon, have shattered the all-encompassing image of the cosmos.

In part, this rupture into disconnected pockets of knowledge is a reflection of the scientific enterprise itself. A unique aspect of science is that its practitioners focus on one aspect of nature, isolating it from all else, measuring everything impinging on and emanating from that fragment. In so doing, scientists gain impressive insights — but only into that separated piece. Scientists, then, learn about nature in bits and pieces, deriving a picture that is both fragmented and unidimensional.

Early in this century, physicists realized that nature is not like an immense jigsaw puzzle whose pieces can be fitted together to provide a comprehensive picture of the whole. Because the pieces behave differently in combination, properties emerge from their interaction that cannot be predicted beforehand. Biologists have been slow to recognize this fact and still believe that if living organisms can be reduced to their most elementary components, a comprehensive picture of life will eventually emerge by their summation.

Once the leisurely activity of gentlemen and aristocrats, science now has enormous national and international consequences. It has grown explosively, especially since World War II, as an endeavor heavily subsidized by national governments for prestige and economic benefit. The rapid increase in the number of scientists (it is said that 90 percent of all scientists who have ever lived are still alive and publishing today) has been accompanied by a corresponding proliferation in disciplines. Thus, a Ph.D. graduate in the early 1960s could simply describe his or her specialty within biology as genetics. Today, one must be much more specific, perhaps referring to oneself as a *Drosophila* (the organism studied) developmental (the specialty) geneticist.

The Scientist As Expert

To the lay public, science is often associated with incomprehensible jargon and a high degree of expertise. Thus, science seems to lie outside the average person's ability to comprehend and evaluate. "Leave it to scientists" is a common sentiment when it comes to determining how new knowledge should be applied.

But who are scientists? What are they like? What drives them? Scientists are seldom accurately portrayed in our popular media. To the public, they are often epitomized by the mad scientists seen in Saturday morning cartoons, or as the noble seekers of truth for human progress. Stripped of passion or romanticized beyond recognition, these images are merely caricatures. Scientists are, above all, human beings with the full range of human foibles and traits.

Given that scientists are human, can't we assume that most scientific endeavors serve the best interests of humankind? The answer is no, for a number of reasons. The most important reason is that the two major users of scientific knowledge today are the military and private industry. Thus, destructive *power* and *profit* are the main engines driving the exploitation of new knowledge. The long-term environmental or social consequences of new scientific applications seldom carry much weight in the face of these priorities. Scientists who are funded by or seek support from the military or private industry have too strong a vested interest in maintaining that support to be objective about their work.

There is another reason that scientists will not always act in the public interest. Within the hierarchy of scientists, maintaining or advancing one's position depends on continual publication and the maintenance of grants. To keep publishing papers and getting grants, scientists must remain glued to the lab bench, where attention is focused on the questions at hand. Under these circumstances, it is difficult to see beyond the barrel of the microscope. Thus, the broader social implications of new discoveries generally lie outside the restricted field of vision of active scientists. Those scientists who do step back to consider the broader ramifications of their research too often lose credibility for ceasing research. Thus, a lay public that is subsidizing the scientific enterprise through research grants, support facilities and tax breaks and that is directly affected by the application of scientific knowledge must be responsible for determining the direction of science and its application.

Controlling the Juggernaut

History informs us that while the benefits of new technologies are immediate and obvious, there are always costs, which are usually hidden and unpredictable a priori. Around the world, governments are urging the scientific community to exploit their discoveries in some practical way. But there is considerable hazard in this headlong rush. While we are regularly assailed by claims of scientific "breakthroughs," most of the information accumulated by scientists represents an incremental increase in our understanding of nature, and new ideas are tentative.

Those who graduated as licensed scientists in the 1960s were at the cutting edge of genetic research. The molecular nature of genes was very well known, and there were elegant models showing chromosome structure and how genetic activity was controlled. Today's students smile at the naiveté of the ideas and models that excited scientists back then, and in view of the knowledge and technical dexterity gained since then, those students are right to be amused. But we must also remember that today's hot idea or theory has a high probability of being tomorrow's outmoded notion. It is only through time that the few important and valid ideas are sifted out, and it is impossible to predict in advance what those will be. So there is great merit in taking time before rushing to apply every new discovery.

We have also lost perspective on how far scientists have come this century. It is true that we have learned more in the past few decades than in all of the preceding years of scientific enquiry. However intoxicating that realization is, we must not lose sight of the distances we still have to travel. There are estimated to be 10 to 30 million species of organisms on this planet. Biologists have only discovered 1.7 million! We know little of the complexity of life on this planet.

Consider the fruit fly, *Drosophila melanogaster*, which thousands of geneticists have studied. For over 80 years, it has been at the center of genetic research. Tens of thousands of scientist-years have been invested in studying *Drosophila*; billions of dollars of research money have been spent on it; four Nobel prizes have been earned by *Drosophila* geneticists — and more will be coming. That effort has paid off. We can do astounding things: We can clone fly cells, and remove and insert genes at will; we can grow flies with 12 legs instead of six, with four wings instead of two, with legs in place of mouths, and with wings sprouting from their eyes.

But after all that investment and knowledge, we still don't know how the fruit fly survives through the winter. We don't understand how a fly's egg transforms itself first into a larva and then into a pupa. We know very little of the mechanisms whereby the flies sense and respond to their surroundings. If scientists learn anything from their studies, it should be how *little*, not how much, we know.

Traditionally, society has responded to new technologies by considering cost and benefit before applying them extensively. When there are perceived health or environmental hazards, risk assessments are carried out beforehand. Thus, it is hoped, we may minimize the deleterious consequences of innovations while enjoying a maximum of benefit. We must understand the limitations of these analytical methods, however. Assessments of health and environmental risks are invariably limited in scale, scope and duration.

The best illustration of these limitations is the oral contraceptive. The benefits of the birth control pill were obvious — profits for the pharmaceutical industry and a powerful method of birth control. An extensive field study was carried out and indicated no harmful side effects. It was only after *millions* of healthy, normal women had used the pill for *years* that we could begin to see significant side effects and discover what to look for. There must be constant accumulation and reassessment of data even after a new technology has been approved.

Perhaps the most troublesome aspect of all technology is the unpredictability of long-term effects in spite of the most careful assessment. Take, for example, biomagnification, the concentration of pesticides at the top of the food chain hundreds of thousands of times above the levels initially sprayed. Biomagnification could not have been anticipated because it was a phenomenon that was only discovered when species of birds began to die off. Another example is nuclear weapons. At the time of Hiroshima, radioactive fallout, holes in ozone layers, electromagnetic pulses of gamma rays and nuclear winter were consequences of nuclear explosions still to be discovered years later.

Thus, as we approach undreamed-of powers to manipulate the very blueprint of life, we must remember the lessons from the history of science and technology. With all of the best intentions, we will still encounter unexpected costs in engineering life. In this book, we hope to provide a few guideposts along the murky path where our new powers are leading us.

CHAPTER 1

THE LEGACY OF GENES

GENETHIC PRINCIPLE
To grasp many of the difficult ethical issues arising from
modern genetics, one must first understand the nature of
genes — their origins, their role in the hereditary processes
of cells and the possibilities for controlling them.

---

*The greatest single achievement of nature to date was
surely the invention of DNA. We have had it from the
very beginning, built into the first cell to emerge, membranes and all, somewhere in the soupy waters of the
cooling planet.*
— Lewis Thomas, in *Discovering DNA* by N. Tiley

*Memory is not what we remember, but that which
remembers us. Memory is a present that never stops
passing.*
                              — Octavio Paz, in "The Curse"

"In the beginning was the gene . . ."

So might a modern biologist's tale — part scientific fact, part
scientific speculation — of the origin of life on earth commence.
A *gene* is life's way of remembering how to perpetuate itself. That
memory is chemical. It is woven into the intricate internal structure of a family of biological molecules, called *nucleic acids*,

found in *chromosomes* and other gene-bearing bodies in organisms ranging from viruses and bacteria to human beings. These nucleic acids are called *deoxyribonucleic acid* — or *DNA* — and *ribonucleic acid* — or *RNA*.

Genes are the vehicle of biological inheritance — the medium through which living things transmit genetic information from one generation to the next. They are the organizing principle by which lifeless raw materials are almost miraculously quickened into living organisms; they are absolutely essential to life. Thus, genes — functional units of self-replicating genetic molecules — must have made their evolutionary debut, many scientists believe, simultaneously with the first faint stirrings of life on our planet.

The Origin of Genes

The first gene is thought to have appeared on the evolutionary scene more than 3.5 billion years ago, in the lifeless, saline seas of the young planet earth. This is a difficult number to grasp except in relative terms. Imagine the complete panorama of the earth's history — about 4.5 billion years — set equal to a 24-hour day (see Figure 1.1). If geological time were displayed on a clock that began ticking the instant our planet was born, the first gene probably emerged in the predawn hours, before 5:00 A.M. Later that morning the first sun-fed photosynthetic cells appeared, followed that afternoon by cells that carry their genes inside a membrane-bound nucleus, and later that evening by the first of many multicellular organisms. The first modern human beings, members of *Homo sapiens*, would not arrive on the evolutionary scene until about the last 30 seconds of this long evolutionary day.

There were no witnesses to record the molecular genesis of this first gene; nor have we recovered any fossil imprint of the momentous event. As a result, we may never know for certain which of a multitude of possible pathways led to this critical transition from nonliving to living — that is, self-replicating — genetic molecule. But modern biology has provided us with more than one scientifically plausible scenario to explain the birth of the primal gene.

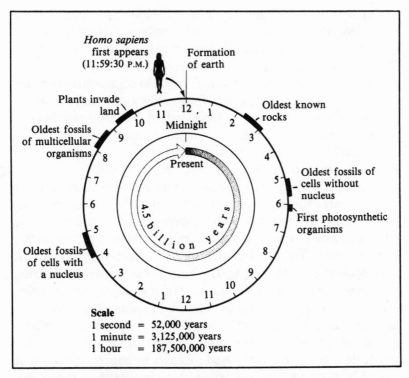

Homo sapiens
first appears
(11:59:30 P.M.)

Formation
of earth

Plants invade
land

Oldest known
rocks

Oldest fossils
of multicellular
organisms

Oldest fossils of
cells without
nucleus

First photosynthetic
organisms

Oldest fossils
of cells with
a nucleus

Present

Midnight

4.5 billion years

Scale
1 second = 52,000 years
1 minute = 3,125,000 years
1 hour = 187,500,000 years

FIGURE 1.1: Evolutionary Clock

Any attempt to reconstruct the origin of genes must deal with a paradox. The first organism had to have some sort of genetic mechanism to ensure the continuity of its kind. Yet even the simplest genetic mechanisms found in modern organisms are fairly complex. How could heredity have been handled in early life forms during the gradual transition from nonlife to life?

Over the years, this apparent riddle has led scientists to propose a variety of possible evolutionary scenarios. Many believe that the first transitional form of life on earth must have been some sort of "protocell" — perhaps a spherical cluster of protein-like molecules capable of crude chemical self-assembly in the absence of any genetic apparatus. Others have suggested that the evolution of a naked genetic apparatus may have preceded the protective cocoon of cytoplasm and membrane found in modern cells. Thus, the transitional life form may have been nothing more than a naked, self-replicating molecule — a simple, primitive genetic mechanism, or protogene, capable of assembling crude replicas from the marine pool of spare parts in which it bathed. And only later, over eons of time, would the ponderously slow process of biological evolution gradually add the additional layers of complexity — such as the insulating apparatus of the cell and eventually multiple cells — to this "primal gene."

Because the latter hypothesis underscores the continuity of genes through the entire history of life, it is also one of the most illuminating. If it is true, the emergence of the first genetic molecule took place in the vast chemical cauldron of the earth's primeval seas. The crust of the young, billion-year-old planet, once molten and aglow from the heat generated by the collapse of a huge cosmic cloud of interstellar dust and gas into the whirling vortex of the new solar system, had by this time cooled and hardened into a tough, textured rind. The earth's atmosphere, yet to be refreshed by the first gusts of oxygen from photosynthetic plants, was composed of gases in concentrations that would be toxic to most modern organisms. Belched from fiery volcanoes, the atmosphere was by this time a mix of methane, ammonia, nitrogen, carbon dioxide and other gases. And it was darkened by clouds that, swollen with moisture, rained down endlessly to feed shallow, expanding, embryonic seas.

At first, these slightly alkaline waters must have been seasoned with only a dilute quantity of relatively simple inorganic molecules, each a cluster of just a few atoms — ammonia, carbon dioxide and hydrogen sulfide, among others — reflecting the primitive composition of the earth and its atmosphere in the utter absence of life. But scientists speculate that a combination of powerful natural forces — sunlight, geothermal heat, radioactivity and lightning — must have provided the critical jolts of energy that made chemical reactions possible. Occasionally, these jolts resulted in the synthesis of larger, more complex molecular structures.

In time, this inexorable energy-dependent march toward greater molecular complexity, known in the scientific world as *chemical evolution*, enriched the seas with a diversity of spontaneously formed organic chemicals. Among them must have been small pools of *nucleotides* and sugars — basic ingredients in modern gene-bearing nucleic acids — and *amino acids* — the building blocks of modern proteins — that circulated in the vast oceanic soup. These ingredients may have interacted with one another spontaneously, in endlessly haphazard collisions fueled by heat, sunlight, lightning and other energy sources. Over periods of millions of years, this trial-and-error process would have mixed these chemical building blocks of modern genetic mechanisms in a staggering number of random combinations. Looking back, one could say that those still-sterile primeval seas were all but simmering with the possibility of new life.

The primal gene may have been nothing more than a rare, eccentric molecular product of such chemical chaos. But it possessed a singular gift — the power of self-replication. It may be impossible to finally establish the identity of life's first genetic molecule. But there is general agreement that it must have been some sort of *polymer* — a chain of smaller, structurally related molecules arranged, often repetitively, like multicolored beads on a string.

Some scientists have suggested that this primal gene must have been a short-chained *polynucleotide* or nucleic acid — a direct ancestor of either DNA or RNA, the stuff of modern genes. The spontaneous synthesis of such molecules would have been a rare event. But once they arose, their structure would have permitted a crude form of self-replication. Other scientists contend that the first gene may have been an unpretentious chain of several amino acids — an ancient precursor of modern protein products, which often contain hundreds of amino acids. Amino acids are thought to have been abundant in the ancient marine environment. But since they are chemically less inclined to copy themselves than nucleic acids, more imaginative schemes are required to explain their first tortuous reproductions.

In either case, a structure with such a repetitive design might have permitted information to be stored in an orderly sequence of chemical subunits or associated bonds. Then, by acting as a mold, or template, this structure could passively select and align free-floating subunits in the vicinity into a corresponding string of molecules. Once the newly assembled polymer thread separated from the parental strand, it also could manufacture reasonable replicas of itself.

Whatever the medium of the first gene — nucleic acid or protein — it must have been endowed with the same three fundamental properties displayed by contemporary genetic materials. First, it must have harbored useful information within its chemically coded structure. Second, it must have possessed the ability to make copies of that information. And third, it must have been capable of changing in ways that would allow it to gradually adjust the content of its message in order to ensure its reproductive success.

With the sudden appearance of such a maverick self-replicating molecule, the engines of biological evolution were set in motion. The primitive genetic mechanism allowed a molecule to produce a wild profusion of chemical offspring, transforming raw marine resources into replicas faster than any other molecule. This self-replication, in turn, permitted its "kind" to command an increasing share of the pool of small-molecule marine resources in its surroundings, at the expense of less innovative competitors.

From the beginning, the process of self-replication was error prone. With each new generation, it introduced an element of chemical imperfection and, therefore, of genetic diversity. With molecular innovation came biological evolution. For not only did this first genetic molecule possess the memory for its own duplication, it also retained a memory of any accidental lapses that fortuitously enhanced its survival. The primal gene thus met the two prerequisites for evolution: possession of a hereditary system and possession of a source of heritable change.

We can only speculate about the sequence of events that gradually led the first self-replicating genetic molecule — nucleic acid or protein — to become integrated into the more complex, clockwork relationship with other nucleic acids and proteins that characterizes the genetic apparatus of organisms today. But whatever the details of the genetic system's evolutionary history, it somehow resulted in the coordination of two vital life processes — the replication of genetic information in nucleic acids and the translation of that information into linkages of amino acids to form useful proteins.

Out of this evolutionary transformation, a provisional *genetic code* — a chemical dictionary that faithfully translated nucleic acid messages into protein messages — appeared. We can imagine that at first the molecular mechanisms that embodied the genetic code must have been primitive and flawed — creating no more than awkward, but tolerable, matches between a minute segment of the genetic material and a handful of amino acid component parts.

But as early life forms began to exploit increasingly complex proteins, built of many amino acids, the genetic code began to require greater precision. At some point, it has been suggested, sudden shifts in the way genes were decoded would simply have been too costly — triggering widespread metabolic malfunctions of time-tested proteins that were likely to prove lethal. As a result, a genetic code that had emerged by chance was suddenly fixed, or "frozen," into the single, relatively rigid code that translates nucleic acid messages into protein products in virtually every species on earth today.

Hypotheses also abound to explain the equally critical encapsulation of genetic materials in a cell. In general, these hypotheses point to the natural tendency of certain molecular raw materials available in the primeval seas to spontaneously align themselves into a membranous globule resembling a microscopic soap bubble. This frail, permeable, spherical layer of molecules could have then provided the embryonic genetic apparatus a measure of seclusion from the surrounding seas. In this way, early precursors to modern cells might have built up increased concentrations of enzymes, structural molecules and other component parts of the complex metabolic machinery that would gradually appear over the eons to come.

Once cells came equipped with a complete genetic apparatus,

stabilized by a fixed genetic code, genetic systems were once again free to diversify. Each species became a potential evolutionary experiment, testing some slight variation of the standard underlying genetic theme — DNA messages copied to RNA intermediates, which, in turn, were translated into protein.

Early Ideas about Heredity

It was not until this century that humans were granted a glimpse, however hazy, into the possible origins of biological inheritance. After all, scientists have only been aware of the existence of hereditary factors they now call genes for a little more than 100 years — and of the molecular nature of the gene for about 25 years. For the rest of our species' history, we have lived in utter ignorance of the lives of genes.

Yet long before the concept of the "gene" crystallized in human consciousness early in this century, human beings felt compelled to search for ways to make sense of at least the most visible evidence of biological inheritance that surrounded them. For they could not help noticing the recurring pattern of reproduction in the natural world by which every form of life seemed to generate new life — "according to its own kind." The keen-eyed agriculturalists among them could not have missed the similarity between successive generations of livestock and crops. Nor was it possible to ignore the sometimes uncanny resemblances between members of one's own immediate family or ancestral lineage. In their state of scientific innocence, early societies were forced to spin elaborate tales to explain the stunning continuity of life they witnessed around them. And they enshrined these beliefs in myth, religion, art and social custom.

To the modern biologist, many of these accounts seem hopelessly at odds with current scientific knowledge about genes, nucleic acids and the process of evolution. But they continue to mirror the deep-seated need of human beings to find confirmation of their own particular worldviews in the natural world and to discover certainty in nature where there is often none.

We cannot know precisely what form the first provisional

hereditary truths of humans took. But we do have fragmentary evidence suggesting that early peoples often displayed a keen interest in the biological inheritance of wild species of animals and plants. By adorning caves with artistic images of wild beasts mating, paleolithic clans, for example, revealed their profound respect for the mysteries underlying the nature of procreation. Ancient Babylonian art demonstrates a fairly sophisticated grasp of the biological separation of sexes in certain species of domestic plants. Babylonian artifacts from 1000 B.C., for instance, often depict priests ritually bearing pinecones dipped in pollen from male date palm trees to the pistillate flowers on female trees to produce fruit. And ancient Greek naturalists, for their part, recorded detailed field observations of the breeding cycles of wild creatures, although gaps in their data sometimes led them to hypothesize fantastic interspecies crosses — a camel mating with a leopard, for instance, as an explanation for the otherwise perplexing animal we call a giraffe.

Early agrarian societies often possessed a firm, if rudimentary, practical knowledge of biological inheritance gleaned from centuries-long experience in domesticating plant and animal species. At first inadvertently and then by design, the earliest agriculturalists, beginning about 10 millennia ago, gradually began to improve their harvests and their stocks by systematically restricting matings to organisms that displayed desirable physical characteristics. In this way, without any knowledge of underlying genetic mechanisms, they successfully selected for crop varieties and domestic breeds by crudely channeling natural evolutionary forces toward useful ends.

Often these societies communicated their hard-won mastery of the principles of artificial selection to future generations through oral or written literature. The early Greeks worshiped Pan, the pastoral hybrid man-goat god of fertility, as one personification of the natural forces they believed to be at play in their agricultural fields and among their flocks. Roman farmers were regularly issued detailed handbooks to guide them in accepted practices of cultivation and animal husbandry. And the

Bible exhorted early Hebrew farmers not to sow their fields "with mingled seed" or to allow their cattle to "gender with diverse kinds."

Humans — past and present — have also been forced to face the stark reality of biological inheritance as it is revealed in the awe-inspiring birth of their own children and in their kinship with ancestors from the distant past. No human society has escaped the need for some sort of system of beliefs to account for the shadowy forces of human heredity that sustain it. Each community has relied on its own distinctive cultural symbols — from mythical tale and sacred text to social ritual — to illuminate these values and to ensure their enduring influence on the lives of its people.

In societies as disparate as ancient Hebrew patriarchies, pre-conquest North American Indian villages and traditional Indian caste systems, for example, power and prestige were often viewed as flowing directly from one's ancestry. In other cultures, one might be allotted anything from a life of misery and servitude to a divinely ordained right to rule, depending on the perceived worthiness of one's forebears.

Not surprisingly, many early beliefs about the nature of human heredity were deeply flawed. Writings composed between A.D. 100 and 300 in India, for example, likened the female in human procreation to a field and the male to a seed. Thus, it was the "united cooperation" of these dissimilar participants that resulted in the generation of new life. And during the fourth century B.C in Greece, Aristotle, for his part, attributed abstract formative powers to a male's semen, which he erroneously believed outweighed any female hereditary contribution.

Blind to the limitations of their flawed and fragmentary visions of human inheritance, early societies were often unable to resist the temptation to try to use this distorted knowledge to shape future human populations. In many cases, such attempts to control the course of human hereditary processes were based on little more than a simple extrapolation of the rudimentary principles of agricultural artificial selection.

Prescribed marriages between individuals of presumed superiority were one way early peoples tried to influence human inheritance. Pharaohs of ancient Egypt, for instance, were known to select their own sisters for wives in order to preserve cherished family characteristics. And in his writings, Plato openly pleaded for human pairings that would foster the most noble physical and moral attributes among members of elite classes in Greek society.

In some cases, strategies to improve on human heredity encouraged the avoidance or even the outright elimination of biologically defective individuals. For example, ancient Hindu sacred texts proscribed marriages to members of families having a history of hemorrhoids, epilepsy, leprosy and other ailments. And the ancient Spartans openly sanctioned infanticide by leaving young children who deviated from the ideal physical type to die by exposure to the elements. But in the absence of any clear grasp of the biological basis of inheritance, such early efforts to improve human populations remained for the most part crude and ineffective. Centuries were to pass before scientists became aware of the mechanisms of biological inheritance and of the ways to manipulate these mechanisms.

Mendel's Genetics

For most of human history, a major obstacle to understanding the nature of biological inheritance lay in the invisibility of sex cells to the unaided eye. Only after the invention of the microscope in the early seventeenth century did these living bridges between generations of multicellular organisms come into view. Not only did the discovery of cells illuminate the underlying mechanisms of sexual reproduction in animals, it also firmly established for the first time, in 1694, the biological basis for sexual reproduction in plants. This breakthrough, in turn, led to systematic studies of plant hybridization and artificial selection, thereby setting the stage for the late-nineteenth-century revelations of the Augustinian monk Gregor Mendel and British naturalist Charles Darwin.

Mendel published the results of his meticulous breeding experiments using hybrid varieties of the common garden pea, *Pisum sativum*, in 1866. But the impact of his work was not fully felt in the scientific world until the early 1900s.

Mendel shattered the popular notion that traits were somehow transmitted along "bloodlines" as diffuse, blood-borne substances that mingled together to create an offspring, in much the same way that streams of molten metal flow together to form an alloy. Essentially, Mendel's experiments followed inheritance of a number of selected physical traits — among them, stem length and seed texture and color — over generations of plants. In the course of his experiments, Mendel discovered that these inherited traits did not blend together as they were passed from parent to offspring. Instead, they seemed to be transmitted as if borne by discrete hereditary "particles" — indivisible genetic factors, borne by both male and female reproductive cells, that somehow maintained their identity while being reshuffled into fresh combinations in descendant organisms. It was not until 1909 that the scientific community came up with an enduring term for Mendel's hypothetical units of biological inheritance — genes.

By monitoring the movement of genes, Mendel uncovered a number of intriguing statistical patterns of inheritance. First, he found that certain categories of genes seemed to exert a stronger, more decisive — or *dominant* — influence on the outward appearance of a plant than did others. Since each pea plant contains a tandem set of genes — one set inherited from each parent — it generally possesses two possible alternative genes, or *alleles*, that determine each Mendelian trait. If the two genes are alike, the plant is called a *homozygote*; if they are dissimilar, it is referred to as a *heterozygote*. Mendel found that a homozygous parental plant with pods bearing smooth seeds invariably gave rise to smooth-seeded offspring, regardless of whether the seed texture of its mate was smooth or wrinkled. That is to say, the gene responsible for the physical characteristic, or *phenotype*, smooth seeds (*S*) was dominant over the weaker, *recessive*, wrinkled-seed trait (*s*) (see Figure 1.2).

Second, Mendel noticed that each gene seemed to be distributed

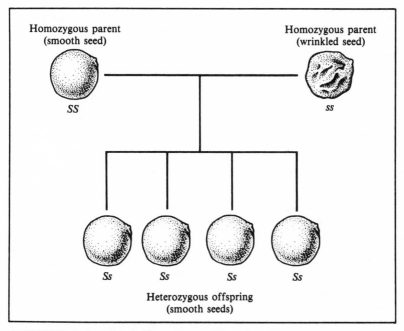

Homozygous parent
(smooth seed)

Homozygous parent
(wrinkled seed)

SS

ss

Ss　　*Ss*　　*Ss*　　*Ss*

Heterozygous offspring
(smooth seeds)

FIGURE 1.2: Mendelian Cross Showing Dominance

independently of its allelic partner. The two genes in each pair of alleles corresponding to a particular trait routinely parted company, or segregated, during the formation of reproductive cells. This meant that one-half of the sperm or egg cell generated by each parent organism contained one allele and one-half contained the other, and that each reproductive cell harbored only one member of each pair of alleles.

Finally, Mendel noted that the separation of each pair of alleles during the formation of reproductive cells took place independently of that of every other pair. Thus, each egg or sperm cell theoretically had a fifty-fifty chance of inheriting one or the other allele in each set of genes that was transmitted in Mendelian fashion.

The surprisingly simple statistical rules that Mendel formulated to describe the pattern by which a small selection of very visible physical traits in pea plants is transmitted marked the beginning of what can only be described as a revolutionary transformation in genetics. In time, Mendel's laws would shatter the subjective, culture-bound belief systems that had shaped all earlier conceptions of heredity and draw attention to the remarkably ordered march of genes from one generation to the next. Now scientists could begin to quantify natural patterns of inheritance and explore the previously hidden behavior of genes in living organisms by tracking their visible manifestations as particular heritable traits. Scientists could also exert much more precise control over the genetic destinies of agricultural plant and animal species. At the same time, Mendel's laws offered tantalizing clues to underlying cellular mechanisms, such as the distribution of chromosomes during cell division, that were later shown to be responsible for the statistical patterns Mendel observed.

The Birth of Eugenics

It is crucial to our later discussions of modern moral issues in genetics to note that from the beginning, societies have always been tempted to apply their knowledge of heredity — however flawed — to the genetic "improvement" of the human species. In some ways, that urge is perfectly understandable. Throughout history, parents have tended to do everything in their power to ensure that they would bear healthy offspring, equipped with socially acceptable characteristics. In primitive societies in which religion or mythology provided compelling explanations for the origins of children, the determinants of their sex or the causes of birth defects and diseases, people naturally turned to ritual, superstitions and other culturally sanctioned means of trying to shape hereditary forces toward desired ends.

Paralleling Mendel's remarkable achievements during the last decades of the nineteenth century — although generally uninfluenced by them — were the efforts of a number of other brilliant scientific minds to design more sophisticated and "scientific" strategies to improve the hereditary qualities of humanity. In 1883, one of the most eminent of these — Francis Galton, an English aristocrat and mathematician who also happened to be a cousin of Charles Darwin — gave this enduring dream of breeding better human beings the name *eugenics* — from the Latin word for "well-born." In his writings, Galton defines eugenics — with a candor seldom encountered among adherents of modern eugenics — as the science of improving the human condition through "judicious matings . . . to give the more suitable races or strains of blood a better chance of prevailing speedily over the less suitable."

Galton used innovative mathematical techniques to search for statistical evidence to support his belief that not only physical characteristics but also intelligence, disposition, artistic talent and other traits were biologically inherited. His notions of "hereditary genius" — a term he used in 1869 as a book title — suggested that class divisions in British society were rooted in innate, hereditary differences in individuals rather than in disproportions in privilege, economic status and power. As we shall see, Galton's ideas would provide the foundations for later reformulations of eugenics designed to reflect — not always faithfully — the insights of twentieth-century genetics.

New Visions of the Gene

Out of Mendel's remarkably insightful studies of genes, carried out in virtual isolation from mainstream science, the fledgling discipline of genetics emerged in the early part of this century. At first, it fed almost exclusively on Mendel's provisional model of the gene as an abstract, particulate unit of biological inheritance. But this model in turn yielded to a long succession of increasingly sophisticated scientific models of the gene.

In a matter of years, for example, it became clear that chromosomes are the gene-bearing structures in a cell. By 1917, T. H. Morgan, an American pioneer of genetics, and others had found that by stalking genes in a species of minute, two-winged fruit flies called *Drosophila melanogaster* — rather than Mendel's peas — they no longer had to postulate imaginary genetic factors. They could begin to point to concrete physical changes in chromosomes that seemed to correspond to Mendel's orderly intergenerational passage of genes. In fact, there was increasing evidence that genes were strung end to end along the length of a chromosome in neat, linear arrays, like beads on a string. As a result of these experiments, the gene was gradually redefined as a single segment of a darkly stained chromosome adrift in the nucleus of a cell.

Later, a student of Morgan's by the name of Hermann J. Muller demonstrated that in the laboratory this beadlike chromosomal gene could be coaxed to transform, or mutate, into a gene coding for a slightly different trait. By bombarding fruit flies with a powerful dose of X-rays, he was able to alter familiar fruit-fly genes controlling wing shape, body bristles and eye color, among other characteristics, into variant forms that produced bizarre traits. These findings established that genes are not the unchanging elements that Mendel had imagined. They are stable factors whose identities can be modified and often rendered more visible for monitoring through techniques of artificially induced mutation.

Eugenics Comes of Age

Meanwhile, by no coincidence, European views about eugenics, nourished by a growing sense of scientific control over the forces of heredity, began to find increasing favor in North America. The American eugenics movement, for example, whose ranks were bolstered by a number of leading geneticists, championed strict governmental policies aimed at promoting what eugenicists perceived to be the genetic health of the nation. Among them were sterilization laws aimed at barring loosely defined "hereditary defectives" from reproducing. Some 30 states passed compulsory sterilization laws that applied to individuals categorized as feebleminded, alcoholic, epileptic, sexually deviant and mentally ill. Even though most such eugenic legislation was never strictly enforced, an estimated 20,000 people had been forcibly sterilized by January 1935 — the majority of them in the state of California. The political climate created by the eugenics movement of this era also encouraged the passage of the U.S. Immigration Act of 1924, which effectively limited the influx of poorly educated immigrants from southern and Eastern Europe on the grounds of their purported genetic inferiority.

It was in Nazi Germany that eugenic theories aimed at purifying the national pedigree were most dramatically put into motion. Building on the "scientific" claims of both the European and the American eugenics movement, as well as its own historical obsessions with racial purity, Germany began to implement eugenic policies shortly after Hitler's rise to power in 1933. During the following year more than 56,000 citizens characterized as "genetically unfit" were sterilized by the German government. Among them were not only the victims of a wide range of mental diseases but also individuals conveniently identified — on the basis of sexual orientation, for example — as "social deviants." In a final expression of his violent contempt for human hereditary differences, Hitler and his henchmen systematically murdered millions of healthy Jews, gypsies and other ethnic and religious minorities — all in the name of eugenics.

The point of this discussion — and of this entire book — is that the applications of genetics, like those of all scientific disciplines, have a dark side as well as the more widely displayed side flooded with optimism's bright light. History confirms that knowledge about heredity has always been vulnerable to exploitation by special-interest groups in society for short-sighted, self-serving, even blatantly cruel ends — often for what seem to be the noblest of motives. Knowing this, we must remain vigilant against future attempts to use genetics to reshape human heredity according to someone's illusory idea of human perfection.

The Double Helix

During World War II, new refinements of gene models continued to surface. In the 1940s, studies of microbial metabolism led to the conclusion that genes act by producing crucial biochemical products. Each gene appeared to be responsible for the cellular production of a single enzyme — a member of a class of highly specific proteins that facilitate, or catalyze, life-sustaining metabolic processes inside cells. Later it became apparent that these protein-producing genes can be partitioned into smaller chromosomal segments, dubbed *cistrons*, that are capable of exchange during cell divisions. The problem is that a gene has many faces; its behavior depends, in part, on the experimental lens through which it is seen.

By far the most compelling vision of the gene to date has arisen from a growing understanding of its fundamental chemical nature. The first clues appeared as early as 1869, three years after the publication of Mendel's monumental publication, when German biochemist Johann Friedrich Miescher extracted a seemingly unimportant substance containing nitrogen and phosphorus from cellular debris found in pus. Since it seemed to arise in the nucleus of a cell, it was called *nuclein*. Not until 1944, with the experiments of Oswald Avery and his colleagues in New York City, did it finally become apparent that nuclein — by then known as deoxyribonucleic acid, or DNA — was, in fact, nothing less than the stuff of which genes were made.

In 1953, James Watson and Francis Crick ushered in a brand new era in modern molecular genetics. They published a brief, but monumental, scientific article that proposed a startling new image of the DNA molecule. They described it as a double helix — a spiraling, two-stranded structure endowed with a stunning symmetry — logical and biological (see Figure 1.3). It was immediately apparent that the Watson-Crick model of DNA offered a plausible chemical basis for DNA self-replication — a molecular attribute that has been essential to biological inheritance since the first ancient stirrings of some anonymous primal gene.

In ways that we will explore in greater detail in chapter 2, the Watson-Crick vision of the double helix launched an intensive exploration of the molecular nature of genes. Within just three years, scientists had unearthed the metabolic pathways by which cells manufacture DNA. By the 1960s, they had cracked the universal genetic code by which trios of nucleotide bases inside DNA's serpentine central core in virtually all species are

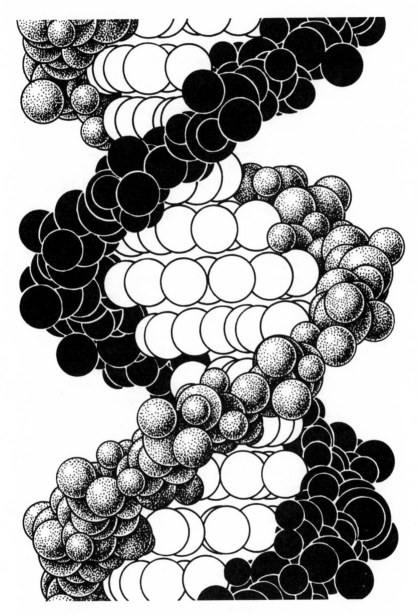

FIGURE 1.3: DNA Double Helix

systematically translated into one or another of 20 possible amino acids. In 1967, a DNA molecule corresponding to a gene was artificially synthesized in the laboratory, and by the early 1970s experiments were underway to snip genes from the genetic molecules of one species and insert them into those of another species. Table 1.1 summarizes these and earlier discoveries in the history of genetics.

TABLE 1.1
Selected Milestones in the History of Genetics and Eugenics

1987	Project to sequence the entire human genome proposed commercially.
1982	Human insulin produced using recombinant DNA techniques.
1982	"Supermice" created by injecting rat growth hormone gene into fertilized mouse eggs.
1981	First human disease is diagnosed prenatally by analyzing DNA.
1977	Scientists develop first techniques to sequence the chemical messages of DNA molecules.
1975	Asilomar Conference considers possible biohazards of recombinant DNA technologies.
1973	First recombinant DNA experiment takes place in which foreign genes are spliced into and function in an organism of another species.
1972	First recombinant DNA molecule created in the laboratory.
1966	DNA's complete genetic code deciphered.
1956	Twenty-three pairs of chromosomes identified in human body cells.
1953	Double-helix structure of DNA proposed.
1940s	A gene is found to code for a single protein.
1943	DNA implicated as the genetic molecule.
1933-45	Nazi Holocaust exterminates six million Jews through its eugenic policy.

1933	Nazi Germany sterilizes 56,244 "hereditary defectives."
1931	Thirty states in U.S. have compulsory sterilization laws.
1927	X-rays found to cause genetic mutations.
1925	Position on a chromosome found to affect a gene's activity.
1924	U.S. Immigration Act limits entry on basis of racial and ethnic origins.
1909	Genetic factors given the name *genes*.
1908	Gene frequencies in Mendelian populations are modeled mathematically.
1887	Scientists discover that reproductive cells constitute a continuous lineage distinct from body cells.
1883	The term *eugenics* is coined.
1871	DNA is isolated from the nucleus of a cell.
1866	Mendel describes fundamental units of inheritance in peas.
1859	Darwin reports his theory of the evolution of species.
1838	Scientists discover that all living organisms are composed of cells.
1677	Animal sperm viewed through a microscope.
1676	Sexual reproduction in plants confirmed.
100-300	Indian metaphorical writings produced on the nature of human reproduction.
323 B.C.	Aristotle speculates on the nature of reproduction and inheritance.
1000 B.C.	Babylonians celebrate the pollination of date palm trees with religious rituals.

The Future of Genetics

Like every earlier vision of the gene, the modern image of the gene as an ordered sequence of DNA is in flux. Over the past decade it has already undergone a number of minor revisions.

For instance, we now know that the genes of higher organisms cannot be defined simply as continuous strings of bases, as they were once thought to be. Instead, the information contained in a gene can be split into fragmentary messages, interrupted by nucleotide sequences that are not expressed. And that information can often be found in multiple copies along a DNA molecule. Moreover, geneticists are finding increasing evidence that the genetic content of an organism is a far less stable store of information than they once imagined. Some genes, it seems, are genetically programmed to be nomadic, allowing them to leap from one cluster of genes to the next.

Genetic research in the years ahead can only continue to expose fresh flaws in our thinking, leading us to ever more sophisticated visions of genes. In the light of the enormous distances we have already traveled in our understanding of inheritance, many of these modifications may turn out to be trivially small. In the course of that long journey, we have left behind an era in which the workings of hereditary were veiled in ignorance, fear and superstition. We have passed through a period of time when almost everything known about biological inheritance depended on tedious empirical observations of the breeding patterns of domestic plants and animals.

Now, at long last, we look beyond the most visible manifestations of genes—physical traits ranging from hair or eye color to the texture of seeds—and gaze at the more revealing landscapes of genetic molecules themselves. This mid-twentieth-century shift in focus has already paid rich dividends. It has enabled geneticists to identify and monitor the movements of individual genes directly, instead of relying exclusively on the indirect evidence of their phenotypic "footprints." At the same time, this shift has given geneticists enormous new powers to intervene in hereditary processes and to manipulate individual genes to satisfy real or perceived human needs.

Until now, the power to determine the fate of individual genes in living things has, with rare exceptions, resided in nature. Evolution tends to rid populations of organisms possessing detrimental genetic traits at a ponderously slow pace. But today we are rapidly assembling the technological tools not only to render quick judgments concerning the "genetic worth" of DNA sequences but also to impose those judgments by modifying the information stored in genetic molecules. Will it be *Homo sapiens* — rather than the time-tested machinery of evolution — that writes the next chapter in the history of genes?

CHAPTER 2

DANCE OF THE GENES

If we had a keen vision and feeling of all ordinary human life, it would be like hearing the grass grow and the squirrel's heart beat, and we should die of that roar which lies on the other side of silence.
— George Eliot, *Middlemarch*

Deep in the interior of almost every cell in your body, inside the membrane-bound nucleus, lie 46 glistening chromosomes — 22 matched pairs, plus two sex chromosomes that normally are structurally identical in females and dissimilar in males (see Figure 2.1). Taken together, these 46 chromosomes represent the standard genetic allotment of most of the approximately 100 trillion cells in your body. The principal exceptions are a fragile lineage of reproductive cells — found in your ovaries if you are female, in your testes if you are male — and a handful of genetic oddities such as muscle cells possessing multiple copies of chromosomes or circulating blood cells that have shed their chromosomes during maturation.

If you could see your own 46 chromosomes magnified a million-fold as they drifted in a minute sea of protoplasm, you would notice that each consists of a single convoluted cord of DNA — the long, linear molecule that contains the genetic instructions

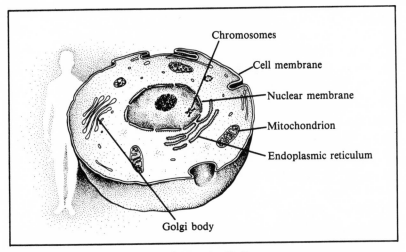

FIGURE 2.1: A Human Somatic Cell

for synthesizing proteins and regulating many cellular processes. Called *chromatin*, this cord is long and elaborately twisted, like a microscopic macramé weaving, around a central supportive rod composed of a type of protein called *histone*.

At first glance, it is difficult to imagine that this tiny quota of genetic material — 46 slender strands of DNA — could possibly encode all of the hereditary information required to orchestrate the life of a cell. But appearances are deceptive. The chromatin contained in each densely packaged chromosome can be unraveled into a single, remarkably fine thread of DNA that can be up to 100,000 times as long as the original chromosome. Magnified a million times, the DNA strand would appear to be about the diameter of ordinary string.

Even without any magnification, these genetic molecules are surprisingly long. If you could stretch all of the DNA contained in the nucleus of this human cell into a single, taut molecular strand, it would extend for about 2.7 meters. If you did this with all of the DNA in all of your 100 trillion cells, it would reach to the moon and back roughly one million times. It is this long, taut thread — a naked DNA double helix — that is the fragile bearer of your genes and the final repository of the genetic information required to animate your every cell.

The Architecture of DNA

Each molecule of DNA consists of two distinct DNA strands joined by weak hydrogen bonds to form a graceful tandem geometric structure (see Figure 2.2). This is the famous double helix discovered by Watson and Crick. Coiling in parallel ascent, like a spiral staircase, each strand of the double helix is composed of four kinds of molecular subunits called nucleotides — each with a distinctively different shape. Each nucleotide contains a sugar, a phosphate, and one of four kinds of nitrogen-containing bases: adenine (A), guanine (G), cytosine (C) and thymine (T). The sugars and phosphates, linked end to end by strong chemical bonds, form the spiraling double spine of the staircase. The bases, projecting inwards from each spiral, are joined near the central axis by weaker chemical bonds to create a soaring flight of stairs bridging the gap between the two strands.

The combined length of all of the DNA stairways contained in this human cell is about three billion base-pair steps, and it is in the orderly sequence of these base pairs making up the molecular staircase that genetic information is ingeniously encoded. DNA's coding system depends on the chemical single-mindedness of the four constituent bases forming the steps of the double helix. Because of their distinctive chemical characteristics, each base insists on joining with only one of its four potential base-pair partners. Thymine combines exclusively with adenine, forming either an adenine-thymine (A-T) or a thymine-adenine (T-A) bond. Cytosine combines exclusively with guanine, forming either a cytosine-guanine (C-G) or a guanine-cytosine (G-C) bond. This base-pairing preference has profound consequences, as we shall see, for the capacity of DNA to serve as a hereditary molecule. For it confers on each DNA strand in the double helix an inherent capacity to store genetic information in the form of a chemical memory bank — the orderly, linear sequences of nucleotide bases.

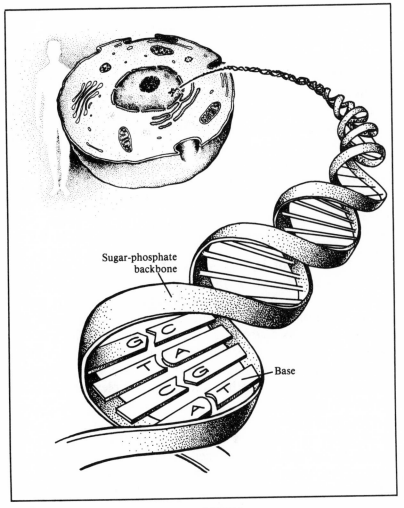

FIGURE 2.2: The Structure of DNA

Thus, the two strands of DNA forming a double helix are complementary to one another. If the strands separate, each is fully capable of recreating the base sequences of the other with the aid of an assortment of specialized enzymes. To do this, each of the two single strands acts like a chemical mold, or template, attracting the complementary nucleotides into position, one at a time, along its length. Then enzymes fuse the nucleotides along their spiral sugar-phosphate spine to form a newly synthesized strand. These two complementary strands of a DNA double helix are related in much the same way as a photographic print and the negative from which it was made. Each harbors the shadowy image of the other and is capable of recreating it.

Since genetic information is stored in a DNA molecule in the order of nucleotide bases, one can imagine "reading" it simply by identifying each base-pair step in ascending order on the stairway. DNA, as we have seen, can reach astonishing lengths. So one would eventually compile a correspondingly long list of base pairs.

The Dynamics of DNA: An Overview

How can such a towering column of base pairs be rendered into messages that are meaningful to the life of a cell? The answer to this question lies in the chemical nature of the gene. A gene can be defined as a stretch of the DNA double helix encoding sufficient genetic information to assemble amino acids into a simple chain of amino acids, or a *polypeptide*. Polypeptides, in turn, can be linked together into more complex, multichained molecules called proteins.

Proteins are critical ingredients in all cells. Some proteins serve as enzymes — molecules that facilitate, or catalyze, specific metabolic processes in the cell. Some, such as the keratin molecules that stiffen skin and hair cells, provide passive structural support. Others — the contractile fibers of muscle cells, for instance — are capable of active movement. Still others serve, in ways largely mysterious to us, as molecular throttles, regulating the activity of genes according to the ever-changing needs of a cell.

While genes may vary considerably in size, a typical gene might include 1,000 base-pair steps in the DNA staircase and about 100 turns in the DNA double helix. Genes, then, can be thought of as a cell's way of partitioning a DNA molecule's long serpentine list of base pairs into units that, expressed as polypeptides, have a specific role in cellular activities.

In the vast majority of species, the messages of genes originate in the chemical structure of DNA molecules. From there, genetic information is conveyed to members of the ribonucleic acid, or RNA, family of molecules that, as fellow nucleic acids, are close chemical kin to DNA. The ingredients in RNA differ only slightly from those of DNA. RNA has ribose rather than deoxyribose sugars (as its name suggests). While it has three of the same nucleotide bases as DNA, it has a base called uracil instead of thymine. Finally, for the most part RNA is a single-stranded molecule, unlike the typically double-stranded DNA molecule. But — except for the fact that uracil, not thymine, bonds with adenine — the crucial complementary base-pairing habits of RNA that encode genetic information remain essentially the same as those of DNA.

RNA in turn transfers the coded instructions to another class of RNA molecules, which direct the synthesis of amino acids into polypeptides and proteins. When the gene's message has reached the end of its one-way passage from nucleic acid to protein, we say that the gene has been *expressed*. At this point, the gene's protein products disperse into the cytoplasm of the cell, joining the proteins coded by other genes. Working together, all these proteins contribute to the myriad metabolic reactions that sustain and give identity to the cell.

So characteristic is the sequence of events by which a gene's message is passed from one genetic molecule to another in a living cell that biologists refer to this pattern of information flow as the *central dogma* of molecular genetics. The term has a rigid, slightly unforgiving ring to it. But the word *dogma* is appropriate; it underscores the remarkably predictable movement of genetic information in all forms of life. According to the central dogma, the information encoded in the DNA sequence of a gene can be transmitted only along two principal pathways (see Figure 2.3). The first route is a one-way cascade from DNA molecules to RNA molecules to protein that is responsible for the constellation of physical characteristics, or phenotype, of a cell expressed by genes. The second route is a circular eddy from one DNA molecule to another as the genes of a mature cell are replicated, or copied, before being transmitted to the cells of the next generation. In the discussions that follow, we will describe the molecular dance of genes as they transmit their precious hereditary information along both of these fundamental genetic pathways.

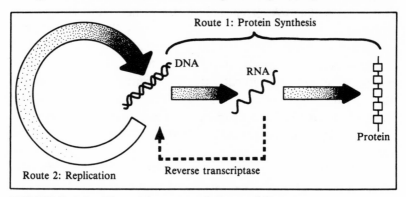

FIGURE 2.3: Flow Diagram of Central Dogma

The importance of the central dogma cannot be overstated. It is testimony to the fundamental biological truth that DNA — not RNA, protein or some other biological molecule previously suspected of playing a hereditary role — is the ultimate source of genetic information in virtually all cells. At the same time, this central dogma shows that the *genotype,* or genetic content, of a cell determines its phenotype, or physical appearance — not, as many early biologists had supposed, the other way around.

It should be parenthetically noted that the central dogma of molecular genetics, however forceful its claim, is not an absolute biological law. The eccentric hereditary habits of a small number of viruses, called *retroviruses*, have inspired a brief corollary to the central dogma's otherwise universal theme. Their genes are inscribed in base sequences of RNA rather than DNA. As a result, during the infection of host cells these so-called RNA viruses must depend on a special enzyme, *reverse transcriptase*, to reverse the usual one-way flow of genetic information from DNA to RNA so that they can rewrite their genes into the self-replicating medium of DNA. Except for a few such genetic oddities, the vast majority of genes on this planet dance to the universal score of the central dogma.

The Dance of Protein Synthesis

To visualize the gene's acrobatic dance from DNA to protein, we will concentrate on the activity of a single gene. Imagine such a gene — a minute fragment of the DNA double helix measuring precisely 1,000 base pairs from end to end, positioned partway up one of the cell's 46 chromosomes, which is now lying uncoiled inside the nucleus of an active human body cell (see Figure 2.4).

Remember, as we describe the steps in the first of two major molecular dances of genes, that what "dances" from DNA to RNA to protein is not the actual physical DNA sequence of the gene itself. Rather, the dancing is done by the gene's fragment of genetic information — its chemically coded prescription for linking corresponding amino acids into a polypeptide chain. The gene itself is neither created, nor dispatched, nor destroyed. It simply conveys copies of its cryptic text to a series of specially equipped messenger molecules, without ever sacrificing its own original script.

Step 1: Transcription

A specialized enzyme assists in the *transcription* of this hypothetical gene's DNA sequence into a mobile, intermediate, working copy of that message in the form of *messenger RNA* (see Figure 2.4). Named *RNA polymerase* for its ability to link, or polymerize, RNA nucleotides into a long, repetitive polynucleotide chain, it is, like all enzymes, itself the protein product of genes. Its task is fourfold: (1) to recognize and bind to the gene; (2) to assist the bases of free-floating RNA nucleotides nearby — previously synthesized elsewhere in the cell — in

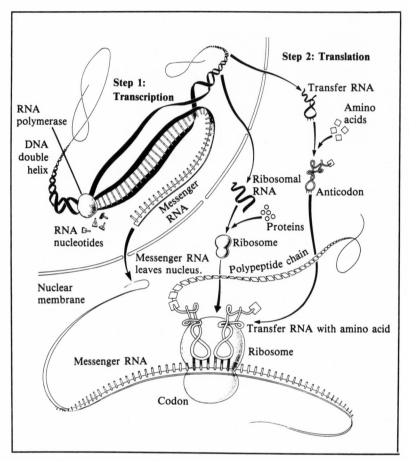

FIGURE 2.4: Protein Synthesis

systematically pairing with their base counterparts on the gene template; (3) to fuse the backbones of the base-paired RNA nucleotides into a linear RNA molecule; and (4) to disconnect itself from the transcribed gene.

The process begins as the enzyme swoops in on the DNA double helix like a bird of prey. The enzyme is equipped to chemically recognize a particular sequence of bases, called a *promoter*, that marks the beginning of the gene's coded message. The enzyme thrusts a shoehornlike extension between the two intertwining strands of DNA, prying them apart. Then it embraces one of them, which will serve as the master copy of the gene during transcription.

Around the gene floats a scattered assortment of molecular spare parts needed for transcription, including a complete inventory of RNA nucleotides, each fitted with a special high-energy chemical bond that will later help weld it into a long RNA chain. The enzyme's distinctive three-dimensional form allows it to simultaneously grip one nucleotide in the DNA chain and one free RNA nucleotide possessing a base that will form a complementary bond with it. The instant the two bases — say, an adenine and a uracil — lock into place, the enzyme shifts, like an old-fashioned automobile jack, one nucleotide notch down the 1,000-nucleotide-long gene.

The process is then repeated for the second base in the gene, then the third, the fourth and so on. Each time, after the enzyme aligns the DNA template base with its RNA mate, it unleashes a brief burst of chemical energy from the high-energy bond attached to an RNA nucleotide, fusing the latest addition to the lengthening chain of RNA nucleotides into place like a link in a steel chain.

The enzyme's ratchetlike progression down the single DNA template strand continues until RNA polymerase has precisely paired each of the gene's 1,000 nucleotide bases to its complementary ribonucleotide partner. As it does, it produces a growing ribbon of newly synthesized RNA that gradually reflects more and more of the gene's distinctive base-sequence memory. When the enzyme finally passes the one-thousandth nucleotide in the gene proper, a special sequence of bases, called the *terminator region*, signals the gene's end. At this point, the enzyme automatically releases its hold on the DNA template strand, as well as its dangling RNA replica. Instantly, the locally distorted strands of DNA spring back into symmetrical double-helix formation. Meanwhile, the newly minted messenger RNA molecule drifts away from the gene, like a fragile ribbon of seaweed, toward its next molecular appointment in the multistaged transmission of genetic information.

This original RNA transcript, like those of most higher organisms, contains the coded messages not only of genes but also of long base sequences that seem to serve as chemical buffer zones between, and even within, actual genes. These intervening sequences, or *introns*, as they are called, are enzymatically snipped from the RNA molecule, and the remaining base sequences — the ingredients for a pure ribbon of genes — are spliced back together to form a working copy of the gene inscribed in the medium of RNA. Once this is done, this messenger RNA molecule, bearing its coded copy of the gene, is propelled through microscopic pores in the nuclear membrane and into the surrounding cytoplasm of the cell.

Step 2: Translation
Here, the mobile messenger RNA's task will be to direct the precisely ordered assembly of amino acids into a specific polypeptide molecule prescribed by the original gene. To do that, the messenger RNA molecule must submit to a process known as *translation* (see Figure 2.4). The messenger RNA's message, written in genetic code, must be deciphered — converted from a linear list of nucleotide bases into one that somehow specifies the entire sequence of constituent amino acids in the gene's polypeptide product.

The cell's decoding apparatus consists primarily of two other varieties of RNA molecules — *ribosomal RNA* and *transfer RNA* — which are themselves products of earlier transcriptions of other genes on one of the cell's 46 chromosomes. Translation of messenger RNA into protein takes place in specialized *organelles* in the cell cytoplasm called *ribosomes* — solid clusters consisting of ribosomal RNA wrapped in protein. These serve as the heavy machinery of protein manufacture, and the cell has thousands of them. Their job is to attach to the ribbon of messenger RNA, convert its cryptic message of nucleotide bases into a message of ordered amino acids and link the requisite amino acids together — one at a time, like so many boxcars in a polypeptide train (see

Figure 2.4). The ribosomes are assisted by a team of much smaller transfer RNA molecules, minute molecules of ribonucleic acid folded into characteristic cloverleaf configurations, which shuttle individual amino acids to ribosomal construction sites.

Translation begins the moment the ribosome recognizes a messenger RNA molecule. At that moment, the boulder-shaped ribosome lets one end of the messenger RNA pass through it, like a strip of film being threaded into a movie projector. Intricately carved crevices in the ribosome ensure that only a small segment of messenger RNA ribbon is exposed at any one time to the swarm of transfer RNA molecules hovering nearby. The segment of messenger RNA molecule framed by the ribosome is precisely three nucleotides long and is known as a *codon*. The exposed codon, which specifies a single amino acid, acts as a sort of sticky molecular landing strip that accommodates a single molecule of transfer RNA at a time.

Each transfer RNA carries a single amino acid and an identifying base triplet, or *anticodon*, that is complementary to one or more messenger RNA base triplets, or codons, specifying that particular amino acid in the polypeptide chain. Thus, a transfer RNA molecule whose anticodon base triplet complements the exposed messenger RNA triplet will suddenly emerge from the cloud of previously synthesized transfer RNA molecules drifting nearby, land on the compact ribosomal landing strip and bond, according to the established RNA base-pairing rules (A-U, C-G), to its codon mate.

As it does, the amino acid mounted on its back slides into position as the first of more than 300 amino acids in the polypeptide chain prescribed by the gene. The messenger RNA template is then advanced precisely three more base pairs ahead through the ribosomal apparatus, until the next codon is in register to be read. This time, the arrival of a complementary transfer RNA molecule not only inserts a second amino acid beside the first but also enzymatically links them together with a peptide bond — the standard chemical coupling device found in proteins.

As each succeeding transfer RNA alights on a newly revealed codon landing strip, it proceeds through an identical series of mechanical steps. First it is attached to the growing polypeptide thread that has begun to extend out from the ribosome like a tail. Next it adds its chemically charged amino acid to that amino acid chain. Then it slides into an adjacent ribosomal holding chamber, thereby clearing the messenger RNA landing strip for the next incoming transfer RNA flight. Finally, the unburdened transfer RNA molecule, stripped of its amino acid cargo, is ejected from the ribosomal holding chamber like a spent cartridge and moves away from the ribosome in search of a fresh amino acid load.

The process of translation proceeds — one codon at at time — until the the ribosome encounters a special termination codon on the messenger RNA molecule. At that point, the ribosome releases the fully translated messenger RNA molecule from its grasp. It also lets go of the newly minted chain of amino acids, or polypeptide, freeing it to twist and turn into its characteristic crumpled three-dimensional configuration, alone or in combination with other polypeptides that will go into making a larger, more elaborate protein molecule. By reading the messenger RNA molecule in sequential three-base increments, or codons, the ribosome and its mobile transfer RNA auxiliary molecules have efficiently decoded the message of the original gene into its final protein product.

The Genetic Code

The key to the translation of a gene's messenger RNA transcript into a polypeptide lies in the orderly system according to which transfer RNA molecules are matched with one or another of the 20 possible amino acids. How, exactly, does the ribosome translate a strip of messenger RNA codons into a corresponding strip of amino acids?

We know that a messenger RNA sequence is written in an "alphabet" made up of just four "letters" — the nucleotide bases cytosine, guanine, adenine and uracil. Furthermore, we know that its genetic message is read in one direction by the ribosome in three-letter "words," called codons, consisting of three adjacent bases. Thus, a messenger RNA molecule is capable, in principle, of coding for a total of 4^3, or 64, three-base combinations. This is more than sufficient to fulfill the requirements of a "dictionary" that assigns at least one unique base triplet to each one of the 20 possible amino acids used to construct polypeptide chains. We call this unwritten dictionary that matches RNA codons with amino acids the genetic code. The complete dictionary, or genetic translation code, can be summarized in tabular form by showing all of the possible RNA codon combinations and their corresponding amino acids (see Table 2.1).

Table 2.1
The Genetic Code
Matching Messenger RNA Codons with Amino Acids

First Base	Second Base				Third Base
	U	C	A	G	
U	UUU Phe	UCU Ser	UAU Tyr	UGU Cys	U
	UUC Phe	UCC Ser	UAC Tyr	UGC Cys	C
	UUA Leu	UCA Ser	UAA STOP	UGA STOP	A
	UUG Leu	UCG Ser	UAG STOP	UGG Typ	G
C	CUU Leu	CCU Pro	CAU His	CGU Arg	U
	CUC Leu	CCC Pro	CAC His	CGC Arg	C
	CUA Leu	CCA Pro	CAA Gln	CGA Arg	A
	CUG Leu	CCG Pro	CAG Gln	CGG Arg	G
A	AUU Ile	ACU Thr	AAU Asn	AGU Ser	U
	AUC Ile	ACC Thr	AAC Asn	AGC Ser	C
	AUA Ile	ACA Thr	AAA Lys	AGA Arg	A
	AUG Met and START	ACG Thr	AAG Lys	AGG Arg	G
G	GUU Val	GCU Ala	GAU Asp	GGU Gly	U
	GUC Val	GCC Ala	GAC Asp	GGC Gly	C
	GUA Val	GCA Ala	GAA Glu	GGA Gly	A
	GUG Val	GCG Ala	GAG Glu	GGG Gly	G

In fact, as you can see, the genetic code relies on 61 of the possible 64 codons in RNA to specify the 20 amino acids. The remaining three codons do not specify amino acids at all. Rather, they serve as punctuation marks in a gene's message. These three base triplets act as *stop codons*, signaling the end of a polypeptide in much the same way that a period signals the end of a sentence. One of the remaining 61 base triplets (AUG) also acts as a *start codon*, in addition to specifying an amino acid. It initiates the translation of RNA into a polypeptide in a manner analogous to the capital letter that marks the beginning of a sentence.

Here is how the genetic code is put into action inside a living cell. Imagine for a moment that the messenger RNA transcript of the gene whose dance we were observing earlier is still in the final stages of translation. If the messenger RNA codon currently exposed on the landing strip of the ribosome is the base sequence uracil-uracil-adenine (UUA), it represents a request for a transfer RNA molecule with a complementary anticodon sequence (AAU) mounted on it like an identifying license plate and carrying the amino acid leucine. If the next messenger RNA codon that slid into the ribosomal reading frame happened to be adenine-uracil-guanine (AUG), it, in turn, would specify a complementary transfer RNA molecule bearing the amino acid methionine. If the next base triplet to be translated turned out to be the codon uracil-guanine-adenine (UGA), it would terminate the translation of the gene's message and signal the end to this brief flurry of protein synthesis.

Since there are 20 amino acids and 61 possible codons, it is apparent that some of the codons specify the same amino acid. This redundancy in the code can be traced to specific biological events during the process of messenger RNA translation. First, a number of transfer RNA molecules are equally adept at transporting the same type of amino acid. Thus, certain amino acids can be ferried to the ribosome by alternative transfer RNA vehicles — some of them bearing a different anticodon; other amino acids are delivered exclusively by one type of transfer RNA.

Second, certain transfer RNA molecules can bond with more than one type of messenger RNA codon. Thus, they introduce an additional element of ambiguity into the genetic code. In the end, only two amino acids — methionine and tryptophan — are specified by unique codons. Others, such as leucine and arginine, are specified by as many as six different codons. The origins and long-term evolutionary significance of these ambiguities in the genetic code remain the subject of intense scientific debate.

The biological significance of the genetic code cannot easily be overemphasized. The relationships between codons and amino acids contained in Table 2.1 apply not just to human genes but, with only a few minor exceptions, to the genes of every species on earth. This means that a gene that has been transcribed into the messenger RNA sequence uracil-uracil-uracil (UUU) will be genetically translated into the amino acid phenylalanine — whether it occurs in a human being, a fruit fly, a redwood tree, or a bacterial cell. In other words, the genes of all organisms are encoded in the same universal genetic language. The revelation that all life is built on a single genetic code confirms our evolutionary unity with other species. At the same time, it has been responsible for igniting a revolution in our ability to analyze and manipulate individual genes.

Repeat Performances of a Gene's Dance
The original DNA-encoded gene we have tracked has by now transmitted a copy of its genetic text DNA to RNA to polypeptide. But it has not necessarily exhausted its usefulness to the cell. Depending on the particular needs of the cell, the entire sequence of steps in this gene's molecular dance — transcription, translation and protein synthesis — may be repeated again.

For, like a phonograph record, a gene's instructions for building a polypeptide can be played over and over again. Thus, the DNA sequence of our original gene may end up being transcribed again later, back in the cell's nucleus where it now lies, to yield fresh batches of messenger RNA transcripts as they are needed in the future. In some genes, the secondhand messenger RNA transcript lingers long enough in the cytoplasm — the rich protoplasmic soup of the cell outside the nucleus — after it has been released from the ribosomal translating machinery to be reread by a succession of other ribosomes. The messenger RNA molecule is simply fed sequentially into a long assembly line of ribosomes, which promptly churn out an equivalent number of identical polypeptide chains, like so many freshly printed newspapers flying off a printing press. In other cases, a gene's output is amplified simply by maintaining multiple copies of its base sequence within the cell's DNA molecules. In this way, repetitive, chemically identical genetic messages can be simultaneously transformed into the same polypeptide formula.

In addition, cells rely on a variety of regulatory mechanisms, most of them poorly understood, to control which gene, among the tens of thousands of genes in a human cell, is active at each moment in time. We do know, for example, that complex cells like ours often harbor special regulatory genes that code for specific molecules that can affect the expression of a gene. It is thought that these control molecules carry out their regulatory missions by interfering with a gene's dance from DNA to RNA to protein at one or more steps along the way. Genes can also be affected by their geographic proximity to other genes. But not all genetic controls reside within the cell itself. The timing and rate of a gene's production of a useful protein product can also be influenced by all sorts of factors external to the cell, ranging from the availability of blood-borne nutrients and changes in temperature to the stimulating effects of specialized protein messengers, such as hormones, synthesized by cells located in other areas of the body.

But, for now, the coded information content of our 1,000-nucleotide-base-long gene has completed its one-way journey and found expression in the highly ordered world of proteins.

The Dance of Replication

The dance of genes, as we have discussed, is not confined simply to the synthesis of proteins during the life of a cell. A gene is ultimately defined by its ability to direct the assembly of amino acids into a polypeptide. For that reason, we began our discussion of the dance of genes with an account of their role in protein synthesis. But in order to transmit the instructions for protein synthesis to other cells, a gene must also possess the ability to replicate, or reproduce itself. And it is DNA *replication* — the process by which genes are copied within the cell — that is central to the role of DNA as the molecule of biological inheritance.

DNA Replication: An Overview

During DNA replication, the DNA double helix unwinds, allowing each of its single strands to serve as a template for the synthesis of a complementary single strand (see Figure 2.5). Relying upon the same rules of complementary base pairing that guide DNA transcription and translation, each strand provides the chemical memory for the synthesis of its missing mate. Then each pair of DNA molecules — one old strand and one new — intertwines to create a complete double helix. Thus, by transmitting one copy of the original double helix to each of two daughter cells, a cell communicates genetic information to subsequent generations.

In some ways, DNA replication bears a strong resemblance to DNA transcription. First, it requires a specialized enzyme, *DNA polymerase* — comparable to RNA polymerase — to catalyze the operation. Second, it requires that the two strands of the DNA double helix separate locally, exposing a base sequence that can serve as a template for a new strand of DNA. Third, the basic ingredients of the new strand — in this case deoxyribonucleotides rather than ribonucleotides — are fitted with special high-energy bonds that will later help fuse them chemically into a long chain.

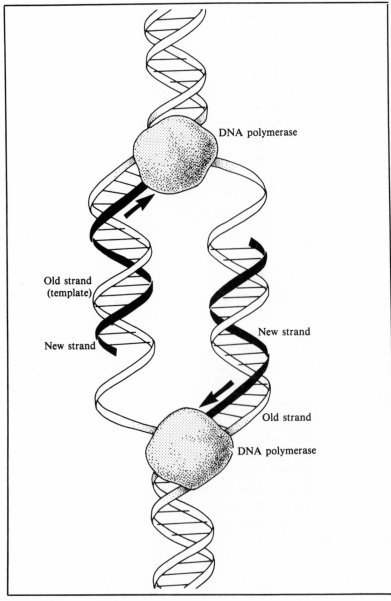

FIGURE 2.5: DNA Replication

But there are also several fundamental differences. Replication of the DNA double helix, unlike transcription, employs both strands simultaneously as templates and, as a result, builds two new DNA molecules, one for each strand. It takes place quickly and accurately on all 46 chromosomes at approximately the same time, rather than piecemeal and erratically according to the changing appetites of the living cell. And upon completion of DNA replication, each newly synthesized strand of nucleic acid remains intertwined with its complementary template strand, instead of disengaging and setting out on further molecular adventures in the cytoplasm of the cell. The end result of replicating a single DNA double helix, therefore, is a pair of double helices — each containing one original DNA strand and one freshly baked one.

The Replication of a Gene
Replication takes place inside the cell nucleus. Triggered by vague chemical tremors that warn of an impending cell division, replication will eventually touch all 46 chromosomes as they lie uncoiled and suspended, like an undersea forest of kelp, in the shallows of the nucleus. It proceeds fitfully; not, as one might imagine, in one continuous crescendo of gene copying from one end of a chromosome to the other. Thousands of localized acts of DNA replication may be going on during the same period of time. These active zones of DNA duplication, known as *replicons*, are usually between 50,000 and 300,000 base pairs long. Oblivious to gene borders, they duplicate long stretches of DNA base sequences within domains that may encompass many genes.

Action begins as a battery of special replication enzymes bombards the DNA double helix in a replicon region that contains our familiar hypothetical gene. The enzymes strike a precise target site at the midpoint of the replicon segment, prying the paired strands apart, unwinding them and then quickly splicing them into two short lengths of RNA to cover the exposed stretch of DNA bases. These RNA patches seem to be necessary to coax a somewhat reluctant DNA polymerase enzyme into action and will be jettisoned as replication gets underway.

A team of two DNA polymerase enzymes now swings into action. Snipping the DNA with surgical precision, both wedge in between the paired strands. Then, positioned at midreplicon and anchored in place by a web of proteins, the two enzymes methodically unravel the mobile double-helix thread of DNA as the double helix is slowly reeled in toward them from two different directions.

The region of separation between the two component strands of the DNA double helix becomes an area of active DNA replication. One of the unzipped DNA strands acts as a template for the synthesis of a new complementary DNA strand. And the DNA polymerase enzyme begins to stitch in new nucleotides, one at a time, out of the nearby pool of spare parts in the nucleus. The other unzipped DNA strand, for now, is ignored. This pattern is exactly mirrored by similar events taking place at the second DNA replication site initiated by the other DNA polymerase enzyme.

It may seem odd that as each stationary enzyme unravels the incoming double helix, it does not instantly spin complementary strands for both of the separated single strands in its wake. If it did, two complete double helices — each with one old strand, one new — would ooze from it like seamless linguini noodles from a pasta machine. Because DNA polymerase can read DNA sequences in only one direction, it can duplicate only one of the two uncoiled strands; the other languishes in single-stranded solitude for a time. Later, a second team of special enzymes moves in to complete the replication of this untouched stretch of DNA. But this time replication takes place not continuously but — according to subtle chemical dictates — in awkward lurching fragments, beginning at the tail end of the replicon.

The Meaning of DNA Replication

When all of this last burst of replicon busywork is done, the most urgent genetic requirement of the approaching cell division has been met; two new daughter threads of DNA have replaced the single one that preceded them. These two complementary strands of DNA are fused together by a centromere to form a single, postreplication chromosome. Each daughter DNA double helix happens to be a molecular hybrid — one weathered DNA fiber spiraling about one newly synthesized one. Except for a few inevitable chemical blemishes arising from errors of manufacture, environmental hazards in the nucleus, or other imperfections of this fragile system, each new double helix stands as an astonishingly accurate replica of the entire sequence of nucleotide bases within the replicon. As a result, every gene in this replicon realm now exists in duplicate. When later events of cell reproduction finally unfold, one exact copy of each such gene will be awarded to each newly emergent daughter cell.

But the same sort of gene replication has been taking place in countless other replicons on this chromosome and on those of the remaining 45 chromosomes, though not in perfect synchrony. Once the work of every replicon on a given chromosome has finished, the fragments of duplicated DNA are joined end to end to form a new chromosome consisting of two identical chromatids. To appreciate the complexity of duplicating the entire genetic content — or *genome* — of a cell, one must envision the tightly orchestrated molecular events we have just witnessed in the microcosm of a single replicon multiplied tens of thousands of times into one sprawling nuclear landscape of replicating DNA. In this context, the dance of one replicating gene, performed on the stage of a single replicon, is just a minor song-and-dance routine within the fantastic festival of a cell's genetic dances.

Or, to use another metaphor, it is as if the replication of genes rested upon the combined labors of tens of thousands of master seamstresses. Each of them is asked to examine one section of a number of fine old silk brocades and to sew an immaculate forgery to match it. As soon as each finishes, she is asked to stitch her handiwork to her neighbor's. When all hands have finally ceased to move, the women find that they are surrounded by 46 sets of shimmering paired lengths of silk fabric. Each bolt in each pair looks identical to an original brocade; yet each is an ingenious blend of textile — ancient and new.

CHAPTER 3

THE DANCES OF THE CHROMOSOMES

What immortal hand or eye
Could frame thy fearful symmetry?
                    — William Blake, *"The Tiger"*

We have just observed, under enormous magnification, the graceful molecular dance of a gene on the stage of a single human cell. During this dance, a gene's chemically coded message is communicated in an orderly fashion to other genetic molecules. It is transcribed from DNA to RNA, which in turn is translated into polypeptide during protein synthesis; it passes from DNA to DNA during replication.

Now, under considerably lower magnification, we are about to watch the ritual dance of entire chromosomes — densely coiled rods of DNA containing thousands of individual genes — as they are distributed during the division of a human cell. Unlike the gene's molecular dance, the chromosome's cellular dances are not confined to the arena of just one membrane-bound cell. Instead, the events of chromosomal dances extend beyond the boundaries of a single parental cell to involve the two daughter cells of the next generation that are its genetic heirs.

Cells reproduce and distribute genetic information to their descendants in a variety of ways. The chromosomal dances we will witness are characteristic of the nucleated cells of higher plants and animals, including ourselves. Simpler but analogous forms of cell division — with corresponding hereditary consequences — occur in bacteria, blue-green algae and other less complex organisms. But we will not focus on these here.

In human body cells, or *somatic cells*, cell division is called *mitosis*. It is the mechanism by which skin cells, muscle cells and other nonreproductive cells perpetuate themselves during normal growth and maintenance. In the development of reproductive cells, or *gametes*, cell division is distinctly different and is called *meiosis*. It is through meiotic cell divisions that human eggs and sperm are created.

The Chromosomal Dance of Mitosis

Each somatic, or body, cell in human beings possesses two complete sets of 23 chromosomes — a total of 46 chromosomes. A cell having two sets of chromosomes is called a *diploid cell*. One set of chromosomes can be traced to the genetic cargo of a father's sperm; the other to that of a mother's egg, or ovum. The sperm and egg each contain half the normal chromosomal quota — just one set of 23 chromosomes — and are called *haploid cells*. When haploid reproductive cells combine during fertilization, they create a diploid cell containing a full set of 46 chromosomes called a *zygote*.

The zygote integrates paternal and maternal information and transmits it, by way of a series of mitotic cell divisions, to every somatic cell in the developing organism. Each gene site on a paternal chromosome is matched by a corresponding gene site on a structurally similar maternal chromosome. Because the genes of the two chromosomes are linked in the same linear order, reflecting a common evolutionary history, the pair of chromosomes is said to be *homologous*. The function of the mitotic dance is to assign one copy of each member of the homologous pair — 46 chromosomes in all — to each of two diploid daughter cells arising from a cell division.

Prelude: Interphase

As the imaginary curtain rises on the chromosomal dance of mitosis, we see the dimly lit stage of the interior of a single cell during the prolonged period of growth and metabolic activity, called *interphase*, that precedes cell division (see Figure 3.1). Cells usually spend most of their lives in this nonreproductive interlude between mitotic divisions. Some, such as mature nerve cells and muscle cells, never divide. They simply live out their lives in interphase, dutifully carrying out their appointed functions in the body but never transferring copies of their genes to progeny cells.

For the moment, the cell's chromosomal dancers are nowhere to be seen. They have not yet begun to materialize from the fibrous swirls of unraveled DNA molecules, or chromatin, contained in the thin, cellophanelike sack of the cell's nucleus. During interphase, many genes etched into these diffuse threads of DNA enter a period of intense activity. They are actively coding for enzymes and other gene products that, by interacting with substances in and around the cell, organize every aspect of cellular life.

But invisible preparations for the chromosomal performance have been in progress long before we arrived. Inside the sheer nuclear envelope, the 46 long, threadlike DNA helices — each corresponding to a single unraveled chromosome — have been silently synthesizing exact molecular replicas of themselves for hours. By the time this crucial process of DNA replication — the gene's second molecular dance described in chapter 2 — has finished, the total quantity of DNA in the nucleus has doubled. Each of the 46 double-helix molecules has become a replicated double helix — identical copies attached at a single point like Siamese twins. And each pair of homologous genes on corresponding chromosomes now exists in duplicate form. As a result, the DNA-laden nucleus before us is packed with twice the normal diploid quota of genetic information, precisely the amount needed to equip two new diploid daughter cells. During the mitotic dance, this genetic information will be distributed equally between the new progeny.

Act 1: Prophase
The first stage in the dance of mitosis, called *prophase*, is the gradual coiling of the fibrous chromatin into the rod-shaped clusters of DNA we call chromosomes (see Figure 3.1). Slowly,

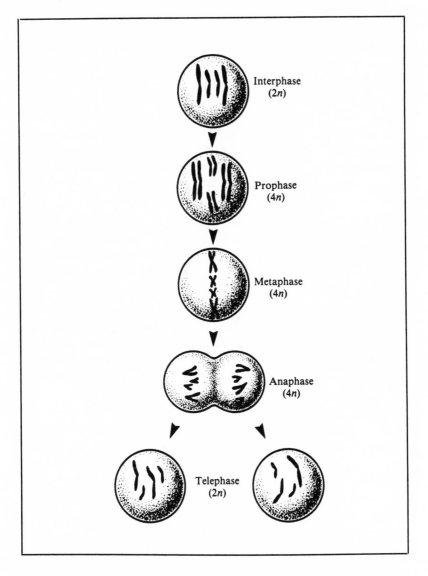

FIGURE 3.1: Mitosis in a Simplified Cell Having Only Two Pairs of Chromosomes (2n = diploid; 4n = twice diploid)

almost imperceptibly, as if to the rhythm of some distant musical score, the transformation begins. Inside the nucleus, the tangle of finely spun duplicate DNA threads seems to spring to life. In response to the signals of a host of genes that act as the choreographers, the threads tremble, twist and turn as they begin to coil methodically around flexible rods of histone protein. Each long twinned strand then weaves itself into a compact, intricate cylindrical macramé of DNA and protein to form a visible chromosome. Gradually, 46 of these ghostly chromosomal figures take shape inside the cell's nucleus like 23 pairs of dancers — the partners closely matched in physical size and stature, as well as in the genetic messages they bear.

The blurred outline of each dancer slowly comes into focus. A chromosome in a resting, nondividing human cell normally consists of a single tightly coiled rope of DNA and protein. But because DNA replication has just occurred in this dividing cell, each chromosome now consists of two identical, coiled, ropelike strands, or *chromatids*, clasped together by a knotlike *centromere* (see Figure 3.2). Thus, a number of the chromosome dancers have what appears to be a limp, four-limbed form. In some, chromatid limbs of equal length cut a symmetrical X-shaped figure. In others, the eccentric location of the centromere may stunt the length of upper or lower limbs, creating figures that look like the letter Y or the letter *V* or some variation of those letters.

Each homologous pair of chromosomes displays remarkable kinship in size, structure and genetic content. Only one pair — the X and Y sex chromosomes, which determine gender — tends to defy this pattern. Human sex chromosomes can appear in two distinct forms. The X chromosome is gangly and resembles its name; the Y, considerably smaller, is little more than a dot. A cell from a female has two physically identical X chromosomes; a cell from a male has a visibly dissimilar pair — one large X and one smaller Y chromosome.

The costumes of this entire troupe of chromosomal dancers are uniformly pale and glistening, and lack any obvious markings that might be used to distinguish one homologous pair of

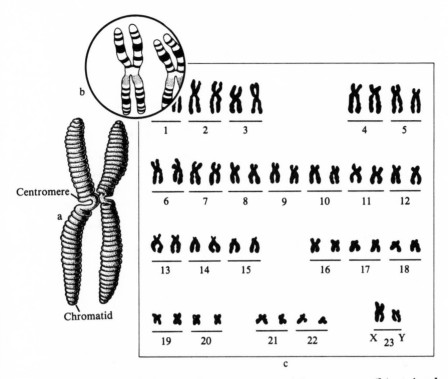

FIGURE 3.2: Human Chromosomes: (a) structure; (b) stained chromosomes; (c) Karyotype of a male

chromosomes from another. But, as we observe the chromosomes in motion, we should remember that each dancer does possess a measure of individuality. If the chromosomes of a cell are treated with special laboratory stains and photographed through a microscope, it is possible to see this cast of performers in a startlingly different light. Such a magnified group portrait of chromosomes is called a *karyotype* (see Figure 3.2). It allows scientists to highlight subtle variations in the chemical composition of chromosomes along their entire length. Staining, in effect, transforms the pale, anonymous natural costume of each chromosomal dancer into a distinctive, zebralike pattern of horizontal stripes. No one knows precisely what these stripes represent, but the shade, size and spacing patterns of these bands confer on each chromosome an individuality that is always present, whether the audience sees it or not.

Act 2: Metaphase

Back on stage, as the 46 chromosomes gradually assume their distinctive, rodlike shapes before our eyes, the entire cell surrounding them undergoes changes of its own. Its fluid contents become more viscous, and the nuclear membrane that surrounds the chromosomes begins to dissolve, mixing nuclear and cytoplasmic soups. Around the same time, the cell begins to erect a spectacular system of protein wires around the chromosomes that will later serve to guide and propel them along a genetically plotted pathway to two separate destinations. Called the spindle apparatus, this system of slender cables is spun by two sets of freshly duplicated organelles called *centrioles* that have migrated to opposite ends, or poles, of the cell's nucleus.

Together, the two sets of centrioles spin a silken bridge of spindle fibers between the two poles of the cell, eventually surrounding the troupe of chromosomes in a cocoon of microscopic parallel fibers. If you were to view this spindle apparatus from the equator of a spherical cell, it would resemble a pair of geometric cones placed base to base. The centrioles would lie at each cone tip, with chromosomes idling midway between them. Viewed from the north or south pole of the cellular globe, the spindle apparatus assumes the dazzlingly symmetrical shape of a sunburst.

The spindle apparatus, as soon becomes apparent, is no passive architectural feat. It is a kinetic sculpture — a fantastic canopy of moving cables. Powered by an assortment of fuel-rich energy molecules in the cell, it delivers two loads of chromosomes to opposite destinations, according to the genetically controlled schedule of mitosis. This process begins after each chromosome sprouts its own set of fibers from its centromere. These minute towlines are connected to the rigging overhead, and the chromosomes are slowly hauled into a tight military formation along the midline, or equatorial plane, between poles. This stage of mitosis is known as *metaphase* (see Figure 3.1).

Act 3: Anaphase

Up to this point in mitosis, chromosomes have seemed to wander more than dance. But now the two complete sets of chromosomes — their centromeres neatly aligned in single file across the center of the cell — begin to sway to a common beat. Lashed firmly to the spindle apparatus, which is in turn anchored to the two poles of the dividing cell, each chromosome is pulled, tortuously at first, toward both poles simultaneously. As spindle fibers appear to strain to the breaking point, each centromere junction joining a pair of chromatids snaps in two. The lengthwise splitting of the centromeres marks the beginning of the third stage of mitosis, known as *anaphase* (see Figure 3.1). During anaphase, the whirring cables of the spindle apparatus overhead drag the freed chromatid twins in opposite directions, one toward each of the two poles of the cell.

This sudden separation of chromatid pairs, throughout the entire flotilla of 46 chromosomes, marks the climax of the mitotic dance. In the minutes that follow, one copy of each chromosome — bearing its load of nucleotide base pairs, codons and genes — is transported toward a region of cytoplasm that will become the nucleus of a new daughter cell. The crucial molecular events of DNA duplication took place hours earlier as chromosomes lay uncoiled and blissfully replicating during interphase. But only now — with the physical separation of the two identical sister chromatids resulting from earlier DNA replication — does the cell actually distribute its bounty of newly synthesized DNA to the two emerging daughter cells. In this way, diploid normality is returned, for a time, to this lineage of cells.

Now that the chromosomes have separated synchronously along the cell's equatorial plane, like some miraculous parting of seas, what remains of mitosis seems largely a formality. Both sets of 46 sister chromatids — each chromatid now a bona fide chromosome in its own right — are laboriously winched by their centromeres toward their respective poles. Upon reaching their destination, the 46 chromosomes — 23 homologous pairs — drift idly in the polar cytoplasm for a time like rafts of logs.

Act 4: Telophase
After the two sets of chromosomes have completed their mirror-image journeys to the two poles, the cell enters the final stage of mitosis known as *telophase* (see Figure 3.1). The parent cell now disassembles its temporary scaffolding of spindle fibers and shrouds each set of chromosomes in a freshly synthesized nuclear sack. Around the same time, paired centrioles at each pole replicate, furnishing each future daughter cell with the requisite two pairs that will one day be needed when each carries out its own mitotic cell division. The flimsy membrane that surrounds the cell then pinches inward along its entire equatorial girdle, squeezing the cell into an hourglass shape. As it cinches the cell in two, a doubly thick length of membrane is laid down over the cytoplasmic channel between the two cellular compartments, and the two daughter cells slip apart as easily as a pair of soap bubbles.

Return to Resting State
Since each cell is now genetically equipped to orchestrate its own metabolic processes, its chromosomes begin to uncoil inside the nucleus in anticipation of a new interphase era of gene transcription, protein synthesis and, eventually, DNA replication. Each daughter cell, its metabolic engines governed by messages from its own diploid quota of genes, will go about its assigned metabolic business — manufacturing hair or hormones or heart muscle — as part of the vast multicellular sea of cells that makes up the human body. Except for those cells, such as circulating red blood cells or brain neurons, that seem to weary of reproduction, a small portion of that diploid quota of genes will in time signal the start of still another in a long series of graceful mitotic cell divisions. For most cells, the mitotic dance of chromosomes ceases only with the death of the organism of which they are a part.

The Meaning of Mitosis

More than anything else, it is the breathtaking synchrony of cellular events and chromosomal dancers during mitosis that elevates this dance above other equally elegant but more mundane biological events that routinely occur during the life of a cell: the dramatic condensation of dense chromosomal rods from wisps of chromatin during prophase; the paradelike precision of chromosome alignment at midcell during metaphase; the sudden, climactic parting of two opposite waves of chromosomes during anaphase; and the final poignant pinching off of parental cell cytoplasm — along with its flotsam of mitochondria, ribosomes and other requisite cell organelles — into two genetically complete daughter cells.

While the life cycles of cells vary, a typical mammalian cell might take 24 hours to progress from one mitotic division to the next (see Figure 3.3). Mitosis itself requires only about one hour out of 24. Interphase — the interlude during which chromosomes lie unraveled in threads of chromatin inside the resting nucleus — consumes the remaining 23 hours. During interphase, genes on the 46 double-helix threads of DNA actively churn out molecules vital to the routine metabolic processes of the resting cell. Thus, using time as a measure, interphase represents the prime of a cell's life; mitosis — despite its central role in heredity — seems almost a diversion.

During that final, fleeting fraction of a somatic cell's life cycle, though, all of the gene-bearing DNA molecules in the interphase cell abandon for a time their ghostly existence as faint fibers of chromatin. Convulsing and coiling, the 46 frail double-helix threads condense into 46 stocky, easily recognizable chromosomes. Then — suddenly visible at the magnification at which we have been observing their performance — the chromosomes move in graceful synchrony, like a troupe of trained dancers, carrying out the genetically choreographed dance of mitosis.

The precision of that choreography should serve as a reminder that the continuity of genes from generation to generation is sustained by a remarkably reliable process that, like most evolutionary processes, which have been going on for tens of millions of years, manages to appear perfectly effortless. In higher plants and animals, the mitotic dance reveals the orderly march of gene-bearing hereditary molecules from one generation of cells to the genetically identical replicas of the next. In bacteria, blue-green algae and other unicellular life forms, less complicated — but equally effective — dances achieve essentially the same hereditary ends.

The Chromosomal Dance of Meiosis

During your lifetime, mitosis will always be the dominant mode of cell reproduction in your body. After all, it was mitosis, always generating pairs of diploid daughter cells from a diploid parent, that created the estimated 100 trillion cells of your body from a single fertilized egg. It gave rise to brain cells, muscle cells, red blood cells, liver cells, bone cells — each equipped, at least for a time, with a nucleus containing the same 46 chromosomes, your lifetime allotment of genes. But that first diploid cell in the long mitotic lineage of cells that fashioned your body arose from the fusion of an egg and a sperm, each endowed with only half the normal number of chromosomes and genes.

During its development, the human body is programmed to reserve a small sanctuary of cells, located in the male testis and female ovaries, where a different form of cell division oversees the distribution of genes. Here ancestral *germ cells*, or reproductive cells — the progenitors of sperm and egg — first arise inside the reproductive organs by ordinary mitosis. But, cued by exquisitely timed genetic signals, each subsequently plunges into the highly specialized process of cell division called meiosis — actually a closely linked pair of divisions — to produce four daughter cells, each harboring one-half the quota of genetic information found in their diploid parent cell. (Oddly enough, in females, only one of these four genetically eccentric haploid cells is destined to survive and become an egg.) In males, these ancestral germ cells are known as *spermatogonia* and undergo meiosis continuously starting at puberty, yielding a steady production of hundreds of billions of haploid sperm cells. In females, ancestral reproductive cells are called *oogonia* and are present in a finite number — far fewer than a male's spermatogonia — at the moment of birth.

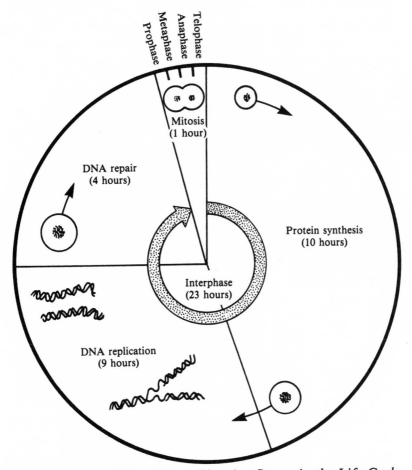

FIGURE 3.3: 24-Hour Clock Showing Stages in the Life Cycle of a Cell

The Meaning of Meiosis

The mathematics of meiosis is relatively straightforward. Diploid somatic cells require 46 chromosomes — two homologous sets of 23 — to fulfill a wide range of specialized body functions. Gametes — eggs and sperm — require precisely half that number, one set of 23 chromosomes, to carry out their singular mission of sexual union to form a diploid fertilized egg. Clearly, if gametes arose by mitosis, each would ferry a full complement of 46 chromosomes to the zygote, bloating it with 92 chromosomes, or two complete sets of human chromosomes. Assuming it managed to survive, a most unlikely prospect, the next generation of mitotically produced gametes would double its 92 chromosomes to 184 following fertilization.

There are moments in evolution when such an exponential rise in one generation's gene count can result not only in survival but even in a decided advantage — in some species of higher plants, for instance. But more often than not, such a monstrous surplus of genes seems to overwhelm the precisely regulated interplay of diploid genes and proves deadly. Thus, in sexually reproducing multicellular organisms, meiosis can be seen as a crucial element in the species' genetic bookkeeping system, balancing accounts to maintain a stable number of chromosomes from one generation to the next.

Meiosis has another crucial hereditary role. Unlike mitosis, it is incapable of rendering perfect genetic copies of the parental genetic molecules. Instead, through a variety of ingenious dance steps, which we will observe in greater detail later, it produces haploid cells endowed with a novel mix of parental genes.

Because the final daughter cells of a meiotic division are usually not genetic replicas of the original parent cell, meiosis might appear to be a less perfect form of genetic transmission than the precision reproduction of mitosis. But by systematically reshuffling genes in each generation, like so many cards in a deck, meiosis also offers at least one distinct advantage over the somewhat more predictable mechanisms and outcomes of mitosis — it produces an endless stream of genetically variant cells. And genetic variability, as you will discover in chapter 4, is the raw material upon which the evolution of all species, including *Homo sapiens*, ultimately depends.

The Setting for Meiosis

To witness the events of meiosis, we must first venture into the cluster of genetically elite germ cells inside the reproductive organs of a human male or female. Let's use a female as our example. Deep in a woman's abdomen, nestled in the bony basket of the pelvis, lies a pair of pale, almond-shaped ovaries that measure about two or three centimeters in length, and have a combined weight of about 70 milligrams. Suspended by a web of ligaments, each small organ rests beside a fringed, funnel-shaped entrance to a slender tube, the oviduct, that leads in turn to the muscular womb, or uterus, of birth. The ovaries, in concert with the pituitary and hypothalamus glands deep inside the woman's brain, secrete streams of highly specific chemical messengers — hormones — that govern the ebb and flow of bloodborne reproductive hormones regulating the menstrual cycle and — following successful fertilization — pregnancy.

But the ovaries are also a storehouse of an estimated 500,000 microscopic egg cells. Haploid products of meiosis, they await the periodic single-egg release, known as ovulation, into the oviduct canal. Only about one in a thousand of them will take that journey, swept along by the action of millions of microscopic cilia bristles protruding from epidermal cells that line the oviduct. Each of these eggs can be seen as a precious genetic time capsule, with the potential to transport a genetically reshuffled haploid version of the woman's genotype toward a timely meeting with a sperm.

The meiotic divisions that created these eggs began, surprisingly enough, during the fifth month of our subject's life in the womb, when germ cells were genetically programmed to begin the process of egg formation, or *oogenesis* (see Figure 3.4). At that instant, each ancestral egg cell initiated a tightly choreographed series of two separate cell divisions that would eventually

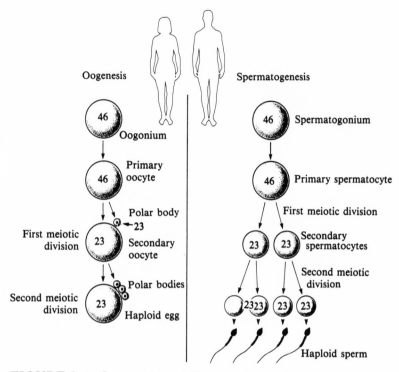

FIGURE 3.4: Oogenesis and Spermatogenesis

parcel out its 46 chromosomes into haploid sets of 23 chromosomes.

Act 1 of Meiosis

At first glance, this meiotic dance, performed in two separate acts, bears a striking resemblance to mitosis (see Figure 3.5). Like the first act of mitosis, act 1 of meiosis is preceded by a prolonged interlude of heightened gene activity during interphase. During this resting stage of the cell, chromosomal DNA uncoils — just as we witnessed in mitosis — into a tangled heap of chromatin threads, while genes actively code for polypeptide products essential to the cell's survival. In time, all 46 chromosomes also undergo replication. By the conclusion of meiotic interphase, the resulting twice-diploid cell is poised — again, just as in mitosis — to parcel out this temporary bounty of DNA to two future daughter cells.

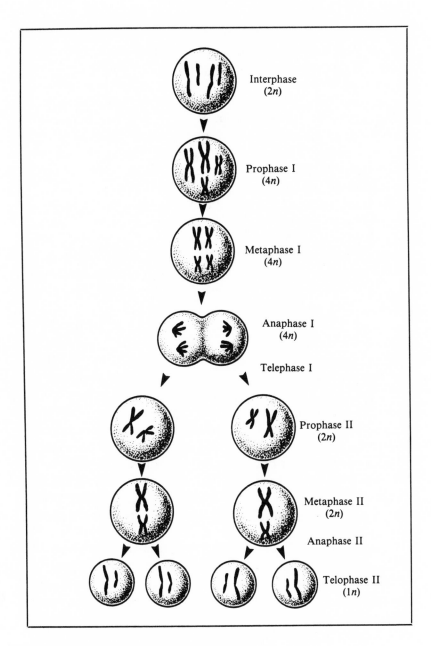

FIGURE 3.5: The Stages of Meiosis

Next, still mimicking mitosis, the parent cell enters prophase I of act 1, during which 23 pairs of homologous chromosomes make their appearance by the condensation of chromatin fibers in the nucleus (see Figure 3.5). The chromosomes of meiosis appear somewhat longer and finer than their mitotic counterparts as they gradually materialize out of the nuclear blue like clouds forming along the edge of an advancing storm. On close inspection, it is apparent that each chromosome is made up of a pair of identical sister chromatids, reflecting the earlier events of DNA replication during interphase.

Yet as the events of prophase I continue to unfold, it becomes clear that meiosis and mitosis follow distinctly different paths. Meiotic prophase becomes increasingly complex, as if it were mitosis gone awry. Oddly, the dividing cell has fabricated no canopy of spindle fibers. And there is no suggestion that the newly formed chromosomes will be towed like barges into the usual equatorial lineup that marks mitotic metaphase.

Instead, each chromosome, anchored at its ends to precise points on the inner lining of the nuclear envelope, appears to be drawn magnetically toward its homologous mate. The two chromosomes — one paternal in origin, the other maternal — approach each other and engage in a long, linear embrace that perfectly aligns the entire DNA sequence of one paternal chromatid with its homologous mate on the adjacent maternal chromatid (see Figure 3.6). A special protein ribbon slips in between the paired chromatids, sealing their gene-to-gene bond like glue. Soon each of the 23 pairs of the two-stranded chromosomes is juxtaposed into a four-stranded parallel repose, known as *synapsis,* that is one of the hallmarks of this stage of meiosis.

If we focus on just one such chromatid quartet, we notice a startling development. Special enzymes are cutting fragments of DNA from one chromatid and splicing them into an adjacent chromatid on a homologous chromosome. Called meiotic *crossing over*, this exchange of corresponding stretches of DNA between two neighboring homologous chromatids appears to be a rather haphazard affair. Along the entire length of a chromosome pair,

Homologous
chromosomes pair.

Chromatids fuse,
break and repair.

After crossing over,
a segment of DNA
has been exchanged
between chromosomes.

FIGURE 3.6: Meiotic Crossing Over of Chromosomes during Prophase I

it usually occurs at least once. But it tends to take place more frequently in chromosomes of greater length. Except for a few special regions where crossing over seems to take place less frequently, these genetic exchanges appear to occur at random sites along the lengths of chromosomes — oblivious to boundaries between genes and swapping DNA segments of variable size.

In the end, two matched segments of DNA fiber originating on separate chromosomes have managed to jump from their respective home-base helices to a nearby homologous one. In doing so, a chromatid of paternal origin has exchanged genes with one of maternal origin. Such a flagrant breach of the genetic "boundary" between two homologous chromosomes is an extremely rare event in mitosis; it is a routine feature of act 1 of meiosis.

Since the exchange of genetic information has occurred between homologous chromatids, strung with virtually identical beads of nucleotide bases, the survival of the organism that develops from this fertilized egg is unlikely to be threatened. But simply by replacing a fraction of the genes on one chromosome with alternative genes from the corresponding chromatid on a homologous chromosome, crossing over shuffles DNA sequences into fresh combinations with each meiotic cell division. It is this natural recombination of genes — along with the endless errors in the messages of genes introduced by natural sources of mutations in the environment — that provides the raw genetic variability essential to evolution.

Once the crossover exchanges are complete, homologous chromatids detach themselves from the nuclear membrane of the cell and relax their hold on each other. But chromatids altered by crossing over remain tethered together at X-shaped crossover junctions, where DNA from one strand of a double helix has detoured to fuse with another.

At this time, the enzymes of recombination rush in to finish their task. They attack any such crossover bridge, shear its two strands, criss-cross the severed ends and splice them back together again. In this way, the wayward stretches of DNA are welded into place and the genetic exchanges resulting from crossing over are rendered permanent. In the process, each altered chromatid is freed from the grip of its crossover companion and slips away from the crossover embrace — bearing the physical scars of meiotic recombination in the mosaic of maternal and paternal base sequences in its DNA.

The first act of meiosis in this dividing *oocyte*, though far from finished, now comes to an unexpected halt. The entire meiotic dance is indefinitely postponed, and the troupe of chromosomal dancers enters what might be called a state of suspended animation that is a common feature of meiosis in women. Poised in the midst of the chromosomal embrace of prophase in act 1, they suddenly halt their meiotic performance. The next phase of meiosis will not begin until the egg is fertilized by a sperm years later.

Meanwhile, the oocyte still has important metabolic business to attend to. Within the sanctuary of an ovary, the developing cell is nurtured by surrounding cells. But it will also continue to express a portion of its own genes — transcribing and translating them into proteins. Not until the midst of a menstrual cycle, after the onset of puberty, will one of the oocyte's chromosomes end their suspended state. The ripened egg will burst from the surface of the ovary and tumble into the dark tunnel of an oviduct. If it is fertilized by a waiting sperm, the chromosomes inside the egg will awaken from their prophase slumber and carry out the final stages of their meiotic dance. If fertilization does not take place, the egg perishes, its meiotic performance forever unfinished.

Signals from the fusion of egg and sperm rekindle act 1 of meiosis. And once again events bear remarkable similarities to those of mitosis. Paired centrioles in the egg cell drift poles apart and spin a scaffolding of fibers. The nuclear membrane disintegrates. Then, as homologous chromosomes — still intertwined following crossing over — begin to drift toward the equatorial plane, the dividing cell enters metaphase I of act 1 (see Figure 3.5). Chromosomes arrange themselves in two parallel ranks of opposing homologue pairs in preparation for the next stage — anaphase I. With chromosomes firmly attached to the mobile spindle apparatus, each member of a pair is tugged, centromere first, toward a different pole. Unlike centromeres in anaphase of mitosis, the centromeres do not split; nor do sister chromatids making up a chromosome separate. Every chromosome simply parts, completely intact, from its homologous mate. Finally, in telophase I, the cytoplasm of the parent cell — along with its two clusters of chromosomes — is partitioned between two new daughter cells.

This parting of ways stands in stark contrast to mitosis in other important ways. Mitosis, you will recall, delivers a complete set of 46 chromosome copies to each daughter cell. In contrast, by randomly assigning only one member of each chromosome pair to a pole, act 1 of meiosis delivers only half that number. Thus, the nucleus of each daughter cell harbors not 46 chromosomes, but 23.

In reducing the number of chromosomes per cell by half, meiosis has succeeded in issuing one copy of each kind of chromosome — and therefore one copy of every gene — to each daughter cell. But it has also accomplished something more. Because the members of each homologous chromosome pair are distributed randomly with respect to every other homologous pair. each daughter cell inherits a mix of paternal and maternal chromosomes — thereby adding new genetic variability into the cell line.

As we have suggested, it is one of the peculiarities of meiosis in female animals, in contrast to males, that only one of these two meiotic daughter cells is permitted to proceed with the dance. Meiosis favors one cell by bequeathing it a larger share of life-sustaining cytoplasm from the parent cell. The other cell is exiled from the parental oocyte as a stunted vestige of a daughter cell, called a *polar body*. Instead of contributing its genes to two new daughter cells, it simply withers and dies.

Act 2 of Meiosis

In many ways, act 2 of meiosis turns out to be little more than a repeat performance of routine mitotic cell division (see Figure 3.5). The 23 chromosomes of the favored meiotic daughter nuclei have replicated long ago during the interphase that preceded act 1. The cell now enters an abbreviated resting stage — interphase II. But during this fleeting interphase, replicas of DNA molecules are not synthesized. This drought in DNA replication during the intermission between act 1 and act 2 of meiosis is absolutely crucial to the final outcome of meiotic cell division. Because this parent cell must now ration out one chromatid from each of the 23 duplicated chromosomes to two future daughter cells, the genetic quota for the next generation of cells will be precisely half that of the current generation. That is, the two daughter cells will be haploid — equipped with one copy of each of the 23 chromosomes — rather than diploid like their parent.

When prophase II of meiosis begins, each chromosome, divorced now from its homologous mate, consists of the same pair of chromatid twins buttoned together by a shared centromere. In metaphase II, the chromosomes line up, beneath a spindle fiber apparatus, in a single row along the equatorial plane. In anaphase II — just as in anaphase of mitosis — each chromosome splits at midcentromere, creating two diverging sets of chromosomes. During mitosis, this ritualized parting produces daughter cells whose genotypes are generally exact photocopies of each other. But because of the genetic reshuffling that has taken place earlier in act 1 of meiosis, the sister chromatids now parting are unlikely to be exact genetic replicas of each other. Crossover exchanges between homologous chromatids may have resulted in DNA sequences that are subtly dissimilar.

The two haploid daughter cells arising from the second cell division of meiosis probably harbor chromosomes that are genetic variants of those in the original diploid parent cell that initiated meiosis. One of these daughter cells will survive to flourish as a bona fide human egg. The other will suffer the fate of the previous generation's sacrificial daughter. It will retreat from center stage, wither and disappear — a second forgotten polar body.

Thus, in women, only one of the four potential products of the complete meiotic cell division proves useful. The other three — a polar body from act 1 capable, in principle, of forming two new cells and a second polar body from act 2 — are discarded as genetic throwaways. This single surviving daughter cell can now merge its singular blend of parental genes with those of a haploid sperm from a male. If this fertilization produces a zygote that successfully becomes implanted in the uterine wall and — through a long series of mitotic divisions — develops into a healthy fetus, this genetic joint venture may prove a resounding success. If not, there will always be another egg cell in the next cycle of ovulation.

Meiosis in Males

We could have just as easily viewed the dance of meiosis within the labyrinths of threadlike tubules inside the male testes, where primary germ cells undergo meiosis to produce hordes of mobile, haploid sperm. In this masculine setting, the meiotic dances of chromosomes are choreographed in essentially the same way that they are in a female (see Figure 3.4). But there are two obvious differences.

First, by the conclusion of the two acts of meiosis, all four potential meiotic daughter cells — not just one — survive to form gametes. But the nature of sexual production demands that male gamete production usually be lavish — an unbroken assembly line extending from puberty to old age. Within the male reproductive organs, 50 million sperm may be manufactured each day — enough, potentially, to populate the earth in about three months. Most of these sperm will, of course, expire without ever fertilizing an egg.

Second, among the chromosomes in a human male's meiotic dance is an eccentric pair of dancers never to be encountered in a female's: the X and Y sex chromosomes that determine human gender. Since females carry only X sex chromosomes, an egg fertilized by an X-bearing sperm will always produce an offspring with an XX genotype — a genetic female. The union of an egg with a Y-bearing sperm will always produce an XY genotype — a genetic male.

In act 1, the 46 chromosomes of an ancestral sperm cell undergo a reshuffling of homologous pairs to create two daughter cells, each possessing 23 chromosomes. In act 2, these 23 chromosomes are divided at their centromeres, generating a complete family of four genetically distinct haploid cells. All four of these final products of meiosis have the potential to mature into compact, torpedo-shaped sperm, capable of fertilizing an egg.

But our focus on meiotic cell division in females does offer a certain aesthetic appeal. In males of our species, the two acts of meiosis, while genetically identical to those performed in the female oocyte, lack her spectacular meiotic dimension of time. The male's chromosomal performance is, in comparison, a furious, nonstop flow of simultaneous meiotic dances that normally continues throughout most of adulthood. In contrast, the female's chromosomal dance is decidedly more patient. Act 1 of the meiotic dance begins, almost hesitantly, in the ovaries of a developing, womb-bound human embryo. Then, before the curtain is lifted on act 2, the entire troupe of chromosomal dancers freezes in midleap and remains motionless in this position until the cell is fertilized by a sperm — an interlude that might last anywhere from a decade to a half-century or more. The performers finish their meiotic dance with a show-stopping flourish, just moments before conception.

Finally, against staggering odds, a single, genetically distinct sperm cell — a survivor of the labyrinthine obstacle course faced by feverishly competing male gametes — fertilizes a single, genetically distinct egg cell — a survivor of the less hectic but no less demanding trials of developing female gametes. And with their fusion, the promise of a new human life takes form.

CHAPTER 4

THE DARWINIAN DANCE: GENES IN POPULATIONS

Evolution is a creative process, in precisely the same sense in which composing a poem or a symphony, carving a statue, . . . [and] painting a picture are creative acts. . . . It renders possible formations of living systems that would otherwise be infinitely improbable. Nothing could be simpler and more ingenious than its mode of operation: gene constellations that fit the environment survive better and reproduce more often than those that fit less well.
— Theodosius Dobzhansky, in *Genetics of the Evolution Process*

In the previous two chapters, we have observed the various dances of genes within cells and individual organisms. These performances included the gene's molecular dance of DNA replication and protein synthesis and its chromosomal dance of mitosis and meiosis. In this chapter, we will explore what we refer to as the gene's Darwinian dance — the evolutionary passage of a gene's message through entire *populations* — groups of individual organisms that belong to the same species and are capable of interbreeding.

The branch of genetics that focuses on the Darwinian dance

of genes is called *population genetics*. Population genetics is concerned with the contribution of individual genes to the genetic constitution of interbreeding populations within a species. It also attempts to monitor changes in the genetic makeup of a population over time periods spanning many generations. That is, it assesses a population's state of genetic *variation* — the genetic differences that exist between individual organisms — at one point in time and contrasts those findings with the state of genetic variation at a later time. It measures the shifting hereditary fortunes of a population according to changes in *gene frequencies* — the relative proportion of alternative forms, or alleles, of a gene at any moment in time. What motivates the population geneticist — and what should capture our interest as we observe the gene's Darwinian dance — is that hereditary changes in populations of organisms over long periods of time are the essence of the evolutionary process. These changes underlie the origin of new species and are the medium through which species are modified over generations of time.

In 1858, in London, Charles Darwin and Alfred Russel Wallace jointly proposed a revolutionary new theory for the origin of species based on *natural selection* — the process by which those individuals of a species equipped with hereditary traits that help them adapt to their surroundings tend to survive and leave more progeny than individuals without those traits. A year later, this theory was formally presented in a landmark book, composed over a period of two decades, entitled *The Origin of Species*.

Neither of these men, it must be remembered, had the faintest notion of what a gene was or even of the basic underlying mechanisms of heredity. In Darwin's era, inherited traits were thought to blend together like paints mixed on an artist's palette. Mendel's revelation of discrete, fixed units of inheritance, distributed from parents to offspring like cards dealt from a deck, was not yet available to weave into their scheme. But because Darwin's theory was founded on meticulous observations of both domestic animals and plants and wild species all over the world, much of it was validated later by the Mendelian concept of the gene. Not until the 1930s did biologists begin to integrate gene theory and evolutionary theory into a synthetic vision that could account for many of the distributions of genes they discovered in the natural world.

Darwin's argument can, in essence, be reduced to three fairly straightforward principles:

1. *The Principle of Variation.* In any natural population of organisms, individuals can be expected to differ from one another in a variety of ways — structurally, functionally and behaviorally.
2. *The Principle of Heredity.* Members of the same species, linked by bonds of biological inheritance, tend to transmit many of their characteristics to their offspring.
3. *The Principle of Selection.* Among this assortment of inherited traits, some inevitably contribute to the survival or reproductive success of an organism more than others.

The evolution of species by natural selection is a natural consequence of these three processes over long periods of time. Organisms harboring hereditary variations that make them more fit in a particular environmental setting — more successful in the evolutionary game of survival and reproduction — are more likely to transmit those traits to successive generations. Organisms that lack such traits are less likely to survive or to leave behind a genetic legacy in the form of progeny. Thus, natural selection gradually modifies the genetic basis of survival and reproduction in all species. In this way, given time, natural selection — in concert with mutation, migration, random events and other evolutionary processes we will discuss that can result in significant shifts in gene frequencies — gradually gives rise to genetically distinct populations, to species and to higher taxonomic groups.

Evolution in Motion: The Case of the Peppered Moth

One way to see the process of natural selection at work is to monitor changes in the frequencies of a single gene over time in a population of living organisms. Every species on earth — including *Homo sapiens* — is shaped to some extent by the gene-filtering effects of natural selection. So, in principle, we should be able to observe the Darwinian dance of genes — those responsible for eye color or blood type, for example, or for some well-known hereditary disease — within populations of human beings. But the staggering number of human genes, the complexity of their interrelationships and the relatively long time span between human generations tend to obscure evolutionary shifts in the fortunes of individual genes.

For this reason, we have chosen to illustrate the genetic basis of evolutionary change through an insect population — European peppered moths belonging to the species *Biston betularia* — rather than a human population. Because peppered moths have been studied intensively in England for more than a century, they are perhaps the best-documented case of natural selection in a multicellular species in all of evolutionary biology.

There is really nothing exceptional about peppered moths. Like most moths, they are nocturnal and possess short, feathery antennae, plump bodies and a pair of powdery, triangular wings that are rather drably colored compared with those of their more flamboyant relatives, the butterflies. For our purposes, it is the wing color of the peppered moth that is of crucial importance.

Originally, biologists noted that in certain British populations of *Biston betularia,* moths came in two basic colors. Most of them had wings and bodies of a pale salt-and-pepper hue that gives the species its common name. But, on rare occasions, moths in a darker, or melanic, shade — a uniform ash-gray — could also be found. As recently as 1848, in a population of peppered moths residing in the area surrounding Manchester, England, the pale-colored moths greatly outnumbered their dark-colored kin. In fact, the pale moths constituted an estimated 99 percent of the local population; the melanic moths made up, at most, a mere 1 percent.

A half-century later, this situation had dramatically reversed — although probably only a few inquisitive naturalists actually took notice of so subtle a change in the local fauna. By 1898, dark moths had come to dominate the peppered moth population near Manchester, making up more than 99 percent of the population; their paler kin made up the rest. What had caused this dramatic shift in the wing color of peppered moths?

Decades of scientific research on peppered moths has revealed that the color changes observed in Manchester moths can be attributed to an exceptionally rapid form of Darwinian natural selection. This process is known as *industrial melanism* — the evolutionary shift within a species toward organisms with more darkly pigmented coloration as a result of nearby industrial development.

To understand the process of industrial melanism, one must first realize that the dark pigmentation in the wings of melanic peppered moths can be traced almost entirely to the effects of a single dominant gene. Two primary alternative genes, or alleles, are responsible for the wing color of a peppered moth: a dark, or melanic, dominant gene — call it *M* — and a pale, recessive gene — call it *m*. Since a normal diploid moth carries two copies of the wing-color gene, one on each of two homologous chromosomes, the alleles for wing color always come in pairs. The visible effect of the dominant *M* gene always eclipses that of its recessive counterpart whenever they appear together. Therefore, there are three possible genetic combinations of wing-color alleles — or genotypes — resulting in two possible physical characteristics — or phenotypes. A moth can be homozygous dominant (*MM*) or heterozygous (*Mm*) and have dark wings, or homozygous recessive (*mm*) and have pale wings.

Second, the last half of the nineteenth century was a time of rapid industrialization in England. Along with industrialization came all sorts of environmental change — including, for the first time in Britain's history, massive quantities of smoke and soot belching from the stacks of coal-burning factories. Air pollution must have affected human populations living around industrial centers such as Manchester — altering the air people breathed, the food they ate, even local weather patterns — often in ways too subtle to measure. But the impact of air pollution on the lives of local peppered moths proved to be especially dramatic. For as soot spewed out of Manchester's factories to settle over the surrounding countryside, it gradually began to blacken trees in nearby forests with a dusting of half-burned debris. It also killed patches of lichen, a pale, mosslike growth of fungi and algae cells that is exceptionally sensitive to airborne pollution and that sheathed the trunks of many large trees in the area.

Tree trunks, it turns out, loom large in a peppered moth's world. The moths rely on tree trunks as convenient places to rest during daylight hours following their usual nocturnal excursions. During the day, as they lie motionless with wings spread against a backdrop of bark, the insects are extremely vulnerable to attacks by keen-eyed predatory birds flying by (see Figure 4.1).

For thousands of generations, moths with pale-colored wings — the *mm* genotype — possessed a built-in camouflage for life in an environment inhabited by trees with pale trunks. Less liable to be spotted by aerial predators, their wing color enhanced their ability to survive to reproductive age and, consequently, to contribute their distinctive recessive *m* genes to the moth gene pool in increasing numbers. In contrast, dark-winged moths — the *MM* and the *Mm* genotypes — remained relatively rare precisely because of their conspicuous sooty phenotype. Through centuries of relentless selective predation by birds, the number of peppered moths displaying the melanic phenotype must have gradually declined. As the melanic phenotype declined, so did the total number of *M* genes in the peppered moth population, since at least one *M* gene was lost with the death of each melanic moth.

Eventually, as the declining number of melanic moths rendered them increasingly difficult targets for predatory birds, a fragile equilibrium must have been reached. As long as environmental conditions strongly favored moths with pale wing color, the vast majority of wing-color genes were *m* genes. Only a small proportion would be *M* genes, all of them harbored by melanic moths that had somehow managed to elude the sharp beaks of predators. By 1848, judging by the proportions of the two color forms observed by naturalists in the local Manchester population,

Normal moth

Melanic moth

FIGURE 4.1: Two Forms of Peppered Moth on a Soot-covered Tree

natural selection had all but eliminated the melanic *M* gene from the peppered moth gene pool.

Then rather suddenly — almost overnight in evolutionary time — air pollution billowing from newly constructed factories began to alter the selective forces acting on *Biston betularia*. No longer were pale moths in evolutionary vogue. Over a period of many generations, pale-winged moths grew more conspicuous in a habitat with steadily blackening tree trunks and were more easily devoured by marauding birds. Dark moths suddenly had a distinct evolutionary advantage. Soon they blended so flawlessly with the bark of sooty, lichenless trees that they began to survive in increasing numbers. And with their sudden, unexpected shift in fortunes came an increase in the number of matings between the melanic survivors.

As a result, each successive generation of peppered moths continued to inherit a greater proportion of *M* alleles than *m* alleles — a natural process of evolutionary bias toward useful genes known as *differential reproduction*. Over a period of years, this process led to a shift in the proportion of *m* and *M* alleles in the moth gene pool. It began to tilt increasingly in favor of the dominant melanic *M* allele, causing the dark melanic phenotype to became statistically more common in the moth population. In sum, a change in an isolated feature in the peppered moth's world led, through the machinery of natural selection, to a corresponding change in the frequency of wing-color genes in the *Biston betularia* gene pool.

Industrial melanism — the evolutionary trend toward darker coloration in species exposed to industrially polluted environments — is not confined to Manchester, England, nor to a single species of moth. It is a recurrent theme in modern evolutionary change around the world. In recent years, similar surges in the proportion of melanic moths and other species of insects and birds have been observed in the United States, Japan, Europe and elsewhere. In each case, there seems to be a clear correlation between the onset of industrialization and the beginning of the melanic color shift. In areas where pollution is most intense, melanic forms tend to predominate; in unaffected environments, light forms hold sway. In England, in fact, a recent improvement in air quality as a result of tighter antipollution regulations has been quickly followed by a measurable increase in the survival rates of pale-colored moths.

Biologists have also been able to confirm critical elements of the theory of industrial melanism through field studies of the predator-prey relationship between birds and moths. It is possible, for instance, after releasing equal numbers of dark and light peppered moths in both polluted and unpolluted forests, to score the success of flycatchers, robins, thrushes or other birds in capturing moths of each color. One can also mark and release a small sample of moths, then recapture a representative fraction of them later — by seducing males to fly into brightly lit traps at night — to gauge which type has fared best in the game of survival. Both sorts of studies, it turns out, have underscored the vital role of wing color in that contest. Compared with their pale kin, melanic moths are, in fact, statistically more likely to be eaten by birds in forests unblackened by pollution and less likely to be eaten in polluted ones.

The case of the peppered moth provides us with a classic illustration of the actual process of natural selection. Not only is it exceptionally well documented, it is also unusually vivid — involving highly visible hereditary traits, relatively simple genetic mechanisms and an evolutionary time frame — measured in decades rather than millennia — that can be witnessed directly during the course of a single human life span. The case of the peppered moth captures the essence of biological evolution — changes in the relative frequencies of genes in populations. It also reveals that the population — the entire community of individuals linked by bonds of mating and parenthood — rather than the individual organism is the principal arena in which the processes of evolutionary change are acted out. Individual organisms have fixed genotypes and finite life spans. Populations, in contrast, are forever in a state of genetic flux and, by persisting from one generation to the next, are potentially immortal.

The case of the peppered moth is also a reminder that the forces of natural selection act on the phenotype — the outward appearance of an organism — rather than on its genotype, or genetic constitution. For it is the final physical manifestation of genes — not the invisible genotype — that ultimately accounts for differences in individual survival and reproduction. In peppered moths, as we have seen, survival in an industrialized world does not finally depend on whether a moth possesses two *M* genes or just one. What matters is whether a moth's genetic constitution, whatever it may be, is expressed in a healthy moth with a pair of darkly pigmented wings.

Thus, evolution is indifferent to precisely which of the possible genotypes is responsible for a particular phenotype. That does not mean, however, that an organism's genotype is evolutionarily insignificant. In fact, it is absolutely crucial. Each time environmental forces select for a specific phenotype, they cannot help but favor those underlying genotypes that code for that trait. Later, when a surviving organism reproduces, only its genotype — the genetic information responsible for phenotype — can be biologically transmitted via genetic molecules to its progeny. Thus, paradoxically, it is a survivor's invisible genotype, culled by environmental forces acting on its visible phenotype, that is the underlying currency of evolutionary change within populations.

Genetic Variation

Natural selection is fueled by genetic variation. Without heritable differences between organisms in a population, the process would simply grind to a halt. If, for example, each peppered moth had the same *mm* genotype as every other — and if the mechanisms of biological inheritance operated with unerring precision — the lineage of moths would unfold each generation in endless genetic uniformity. This genetic uniformity could only result in a population of monotonously pale-winged moths. In a preindustrial era, the pale-winged trait would serve moth survival well — as long as the surroundings remained stable. But a shift in environmental conditions, such as a sudden influx of airborne pollutants, that would normally select against a prevailing trait would not be able to do so in the absence of a pool of genetic options. Instead, the altered conditions could only place upper limits on the size of the moth population or possibly lead to its extinction. Since the proportions of genes in the moth gene pool would remain unchanged, no natural selection could occur.

Nature has two principal means by which it introduces genetic variation into organisms such as peppered moths or human beings. The first is genetic mutation — random changes in the genetic messages of genes or chromosomes. Mutation is the

wellspring from which all raw genetic differences arise, and mutation alone is capable of creating new alleles in a species' gene
pool. The second means is genetic recombination — the continuous reshuffling of genes into new and unpredictable combinations through the chromosomal dance of meiosis. Recombination, as you may recall, is confined to sexually reproducing
species, which rely on meiotic cell division to distribute a mix
of paternal and maternal genes to egg and sperm cells, and thus
into the gene pool of the subsequent generation. The total number
of mathematically possible gene combinations in sexually
reproducing species is staggering. Yet the genotypes of living
members of such a population generally represent only a tiny fraction of those possible combinations. This natural reserve of
genetic variation is tapped each generation by meiotic processes
that assign fresh combinations of chromosomes and genes to
reproductive cells. Unlike mutation, meiotic recombination introduces genetic variation into a population not by forging new
alleles but by redistributing existing ones.

No one knows for certain exactly how or when the first gene
coding for dark wing color in peppered moths appeared. It probably surfaced suddenly — perhaps more than once — as a rare
mutant form of an existing gene in some forgotten population
of moths in *Biston betularia's* distant evolutionary past. There
are a number of ways this ancient mutational event might have
occurred. It could have been nothing more than the substitution
of one nucleotide base for another in the long DNA sequence
that made up the original gene — a misspelling of a single letter
in a gene's script. It could as easily have arisen from the accidental
addition or removal of one nucleotide in that script — resulting
in a far greater disruption of the coded message. For by altering
the total length of a gene's linear text, this sort of event — known
as a frameshift mutation — would stir a wave of nucleotide-base
spelling errors that would affect the entire DNA sequence
downstream from the mistake. The crucial mutational event could
also have been any one of a variety of more monumental rumbl-
ings in the structure of the DNA double helix, leading to unex-

pected breakages and reunions between neighboring chromosomes. These chromosomal mutations would be capable of altering a gene's message by juggling entire sections of a DNA molecule, thereby causing a gene to duplicate, rotate 180 degrees, resurface on another chromosomal raft or even totally disappear.

The original mutation might have been triggered by a variety of forces — all part of a peppered moth's natural surroundings. It could have been caused by the sun's rays or radioactivity from minerals in the soil or an assortment of physical or chemical factors too faint to identify. Regardless of the cause of this rare, harmless mutation, once the modified gene was transmitted to the next generation of moths, resulting in viable dark-winged progeny, the M gene had made its debut in the peppered moth gene pool.

The vast majority of mutations are injurious to the highly ordered instructions encoded in genes. Random change adds an element of chaos — genetic noise — to the information load of a gene. The genetic mistake is quickly magnified as the gene is expressed, often repeatedly, during protein synthesis. And just as the musical score of a symphony is unlikely to be improved by random alterations to its notes, a mutated gene is unlikely to lead to a new, improved version of a complex, often vital protein. An error in even a single amino acid can so distort the delicate three-dimensional design of the molecule that it will twist, fold or collapse — disturbing, even destroying, its capacity to carry out designated cell functions.

Only on rare occasions does a newly mutated gene modify a protein in ways that do not damage its finely tuned metabolic role, honed by countless years of evolution, in a cell. The mutant M gene must have been just such an exception: a one-in-a-million mutational event that happened to trigger increased levels of dark pigment in the wings of a moth, without harming the health of the insect that possessed the novel allele. Even more rarely does a mutation actually add to the evolutionary fitness of a gene. Following the appearance of the new mutant M gene in the peppered moth gene pool during preindustrial times, a fraction of the descendants of the first melanic moth must have managed to survive, despite the temporary evolutionary burden of dark coloration. For only with the persistence of the maladaptive gene in the moth population, or with the rare recurrence of a similar mutational event, would the maverick *M* gene have been available to the species when industrial pollution finally gave it the chance to prove its latent evolutionary worth.

Chance and Migration

Industrial melanism in the peppered moth represents a clear case of natural selection in motion. But natural selection is not the only force that can affect a gene's evolutionary dance by causing changes in the frequencies of genes. Mutation alone, as we have seen, creates new alleles in populations, and this genetic variability, along with that generated secondarily by meiotic recombination, is a vital component of the machinery of natural selection. In the absence of natural selection, the process of genetic mutation generally represents a neglible force in shaping the evolutionary histories of species. Nonetheless, because random mutations themselves cause slight changes in the gene frequencies of a population — regardless of their contribution to the fitness of an organism — they represent a legitimate, if minor, evolutionary force in their own right.

In a sufficiently small population, the evolutionary fortunes of a gene can fluctuate wildly according to the whims of chance alone — a passive process of evolutionary change known as *genetic drift*. In this process, a statistical bias favoring one or another allele is introduced into a gene pool not by the clash between hereditary traits and a fickle environment, eliminating genotypes through death or reproductive failure, but by random events. The contribution of that genetic option to an organism's survival or reproductive success is overshadowed by chance. Just as an unexpected ratio of heads to tails is far more probable after 10 tosses of a coin than after 100 tosses, so is the impact of mutations, meiotic reshufflings of genes and other random events greater in a small, interbreeding population than in a large one.

Small statistical ripples in gene frequency that are likely to dissipate in a large gene pool may become tidal waves of change in a small one. As a result, the statistical laws of probability alone can cause the disappearance of a once-favored gene or thrust a harmless alternative into evolutionary prominence — with little regard to the winnowing process of natural selection.

Migration, the sudden influx of variant genes from a genetically related but geographically distinct population, can also lead to changes in gene frequencies. Matings between migrant and resident organisms produce hybrid offspring — effectively injecting new genetic variability into the local gene pool. The original source of this variability, it must be remembered, was still mutation — random events that occurred earlier in the migrant population.

Forecasting the Fate of a Gene

It is one thing to visualize a gene's evolutionary dance in a population of peppered moths and quite another to describe it in precise mathematical terms. One of the theoretical pillars of modern population genetics is the Hardy-Weinberg equilibrium, a mathematical equation named in honor of the British and German scientists who developed it in the early 1900s in an attempt to compute the changing fortunes of a gene over generations of time. The equation uses the frequencies of a gene's alternative alleles in an initial population to calculate their future frequencies in each subsequent generation. But the predictions turn out to be reliable only under a strict set of preconditions that are seldom encountered in nature.

The Hardy-Weinberg equations set out to predict mathematically the point at which the genetic composition of a population will reach a dynamic balance, or equilibrium (see Table 4.1). In this state, gene frequencies, driven only by hereditary processes, stabilize and remain constant from one generation to the next. In the absence of any change in the relative proportions of genes, evolution comes to a halt. This means that the lives of genes in an idealized Hardy-Weinberg world bear only a faint resemblance to genetic reality. The fundamental equation is one of the few tools available to geneticists trying to anticipate the future frequencies of genes in populations. Yet it is able to forecast these gene frequencies only by removing all influences except biological inheritance — genes endlessly creating exact facsimiles of themselves — and the laws of probability — the random reshuffling of genes during meiosis.

TABLE 4.1
A Sample Calculation Using Hardy-Weinberg Equilibrium

All eukaryotes possess two copies of every gene that may be passed on to subsequent generations. So for each gene, the sum of all of those pairs carried by individuals may be added up to give a profile of the population of that gene. We can represent those two genes in the moths as p (M) + $q(m)$, where p is the incidence or frequency of the M allele in the population and q is the frequency of m. So in any mating pair, the probability of having offspring of a particular genotype is calculated by multiplying the two together as:

$$[p(M) + q(m)] \times [p(M) + q(m)] = 1$$

or

$$[p(M) + q(m)]^2 = 1$$

This is the Hardy-Weinberg equation. It you multiply this equation out, it becomes:

$$p^2(MM) + 2pq\ (Mm) + q^2\ (mm) = 1$$

In other words, the chance of having a homozygous MM moth is p^2, of an Mm heterozygote is 2 pq and an mm homoygote is q^2.

To compute the gene frequencies of M and m alleles in future generations, all one needs to know is the initial frequencies of the two alternative alleles in a parental population of moths.

Consider a preindustrial moth gene pool containing a total of 100 wing-color genes — 10 of them melanic M genes (p = 0.1); the rest, 90 (q = 0.9), the common m gene.

1% *MM* moths (dark-winged homozygotes)
(since $p^2 = 0.1 \times 0.1 = 0.01 = 1\%$)

18% *Mm* moths (dark-winged heterozygotes)
(since $2pq = 2 \times 0.1 \times 0.9 = 0.18 = 18\%$)

81% *mm* moths (pale-winged homozygotes)
(since $q^2 = 0.9 \times 0.9 = 0.81 = 81\%$)

For Hardy-Weinberg computations to remain valid, the population under study must consist of an infinite number of fertile males and females that meet, mate and procreate with all of the randomness of a lottery draw. However mathematically appealing this premise, it rarely occurs in nature — and certainly not in human populations. Human liaisons, for example, in which skin color, height, body type or other physical attributes play a role, are simply not encompassed by the rigid Hardy-Weinberg worldview. The trait being followed must also be traceable to a single Mendelian gene, possessing a limited number of alternative forms; *polygenic* traits — those characteristics that arise from the effects of more than one gene — cannot be mathematically tracked. Perhaps most limiting of all of the premises of the Hardy-Weinberg equilibrium, the species in question must be insulated from the effects of natural selection, mutation, migration and genetic drift — the very processes that fuel the machinery of evolutionary change.

Even with these agonizing limitations, Hardy-Weinberg calculations can still prove extremely useful. For instance, one can compute the expected changes in the frequency of genes in populations under restricted theoretical conditions. Then one can compare these findings with gene frequencies measured in nature. Divergences between the expected and actual frequencies of a gene sometimes offer valuable clues about possible evolutionary forces that might be at work in a particular wild population of animals. In addition, one can also try to improve on the basic Hardy Weinberg equation by incorporating simple mathematical formulas that introduce at least crude approximations of such real-world influences as natural selection, mutation, migration and genetic drift.

Vagaries of the Darwinian Dance

The point is that no scientific model — from Hardy-Weinberg equations to experimental computer programs — has yet come close to mirroring the subtleties of a gene's Darwinian dance through time. There are simply too many variables affecting genes for geneticists to monitor. In the first place, many genes cannot be tracked as visible single-gene Mendelian traits that progress through a population with the regularity of a moth's wing color, for example. Second, a gene is often influenced by its neighbors, strung beadlike along the length of a chromosome, and often has a tendency to remain attached to them during its tumultuous passage through meiotic cell divisions — a phenomenon geneticists call *gene linkage*. Third, genes influence one another in ways that geneticists have only begun to grasp. For example, genes whose disparate geographic locations on different chromosomes might suggest, at first glance, that they are completely unrelated can turn out to influence one another in exquisitely coordinated ways. Geneticists have even identified teams of interacting genes, collectively called *supergenes*, with biochemical roles so seamlessly interwoven that they give the appearance of being transmitted as a single Mendelian gene.

Other genes can prove perplexing because they do not reveal a neat pair of Mendelian alleles for that site. Instead, certain genes, such as those coding for the component parts of the blood protein hemoglobin, can turn out to have so many alternative forms, or *polymorphisms*, that it becomes a major challenge to try to catalog alleles. This variation might be conveniently expressed in physical characteristics of an organism that are visible to an observer. But often it can surface as minuscule chemical changes that can only be detected in DNA molecules themselves — ones that can be inherited but are not revealed by a detectable phenotypic trait.

To further complicate matters, the alternative alleles for a gene do not always interact in textbook expressions of dominance and recessiveness. A pair of variant genes can, for instance, display what is called *incomplete dominance* — combining in heterozygous offspring to create phenotypes that are an intermediate blend of homozygous extremes. Crossing a carnation bearing red flowers with one bearing white flowers, for instance, may result in a carnation bearing pink flowers. Other gene pairs exhibit *codominance* — each producing its own distinctive protein product in a heterozygote organism, thereby expressing both genetic traits simultaneously. In humans, for example, *MN* blood groups, coding for highly specific molecules on the surface of red blood cells, exhibit codominant inheritance.

Even the wing-color genes of peppered moths, it turns out, do not strictly abide by the rigid Mendelian laws of inheritance that humans have tended to impose on them. Although the process of industrial melanism in *Biston betularia* has generally been ascribed to the activity of a single dominant *M* gene, there is increasing evidence that the genetic basis of melanism may be more complex than it first seemed. Recent observations of intermediate phenotypes — moths possessing wings in various shades of gray — suggest that more than two alleles may be involved at the site of the wing-color gene. Furthermore, there are indications that the fate of other, still unidentified genes elsewhere in the moth genome may be linked to those of the wing-color alleles. If so, a portion of the evolutionary fitness attributed to a moth's melanic genotype may be due not only to the camouflage provided by darkly pigmented wings but also to other secondary physical or biochemical traits transmitted in synchrony with the *M* gene.

When the dynamics of genes in relatively simple natural populations sometimes fail to match our textbook population models of genes, it is easy to feel overwhelmed by the sheer complexity of a gene's evolutionary dance. How can we hope to come to terms with the subtleties of evolutionary change as it is acted out by individual human genes — tens of thousands of separate genes diffusing through populations totaling billions of human beings over vast periods of geologic time?

But this question need not disturb us. The history of evolution on earth is richly textured with broader genetic patterns that scientists and nonscientists alike can appreciate and explore. One of them is the startling irreversibility of evolution. No species ever retraces its evolutionary footsteps. At any given moment, the combination of a species' genetic composition and the state of its surroundings is unique. This means that the evolutionary lineage of each species, fueled by its own peculiar mutations and endlessly changing surroundings, is also unique. As each species' constellation of genes engages in its own Darwinian dance — encountering transient evolutionary successes and failures, or even extinction — its future can never be precisely predicted. Thus, in spite of our best efforts to capture the nature of evolving genes in the symbols of mathematics or language, we will never be able to predict the precise future course of evolution or to reliably anticipate the evolutionary consequences of releasing novel, genetically engineered life forms into nature.

Finally, it is important to remember that no single perspective on genes — molecular (including DNA transcription, translation and replication), chromosomal (including mitosis and meiosis) or evolutionary (including natural selection) can be considered complete in itself. In order to discuss genes, we have been forced to observe their movements within each of these three artificially defined realms of gene activity. Each of these three vantage points from which we have watched the dances of genes allows us to catch a fleeting glimpse of one dimension of gene behavior without becoming dazzled by the true multidimensional complexity of a gene's world.

At the same time, we must not allow the artificial simplicity imposed by these three genetic performances of genes to deceive us into believing that they fully capture the essence of a gene. Nature itself imposes no such arbitrary boundaries on gene play. In the real world, all three genetic dances — molecular, chromosomal and evolutionary — take place simultaneously on a single, staggeringly complex global stage. To even begin to sense the elusive reality of genes, we must turn from textbook descrip-. tions to the human imagination. The challenge is for each of us to try to somehow superimpose the diverse roles of genes — from DNA replication to the shaping of species — into a seamless whole that, for the moment at least, lies beyond the scientific description. Even with such a creative vision of genes in our minds, we are still entitled to wonder how many additional dimensions of gene activity have yet to be discovered.

CHAPTER 5

RECOMBINANT DNA: THE NEW CHOREOGRAPHY

Mythology is full of hybrid creatures such as the Sphinx, the Minotaur and the Chimera, but the real world is not. It is populated by organisms that have been shaped not by the union of characteristics derived from very dissimilar organisms but by evolution within species that retain their basic identity generation after generation.
— Stanley Cohen, in "The Manipulation of Genes," *Scientific American*

Long before the science of genetics granted us the stunning visions of genes in motion that we have just witnessed, humans had devised a variety of ingenious ways to intervene in natural hereditary processes in order to alter the genetic constitution of organisms. This ancient urge to choreograph the natural dances of genes to satisfy real or perceived human needs can be considered a form of *genetic engineering* in its broadest sense.

But until very recently in human history, our capacity to engineer genes has been almost embarrassingly limited. For the most part, our attempts to intervene in hereditary processes have been confined to artificial selection — the painstakingly slow process of breeding domestic plants or animals of the same species in order to select for a relatively small number of visible, economically important genetic traits. The repertoire of techniques available for this ancient tradition of genetic engineering remains relatively limited — ranging from making simple crosses between plants and monitoring livestock pedigrees to bombarding chromosomes with X-rays or chemicals to generate useful genetic mutations. Moreover, virtually all of our vaunted experience as genetic engineers — at least 99 percent of it — took place in utter ignorance of the biological basis of what we were doing. Only in the last century have we had even the remotest idea of the hereditary processes we were really "engineering."

In this book, we offer a stricter, more scientifically precise definition of genetic engineering: the use of new and revolutionary laboratory techniques, representing a synthesis of molecular genetics, biochemistry and microbiology, to modify the genetic makeup of cells and organisms through the manipulation of individual genes. Contemporary genetic engineering differs from traditional efforts to shape hereditary processes in its startling new ability to precisely integrate DNA sequences from different organisms and even different species to create hybrid DNA molecules that we refer to as recombinant DNA.

This difference is apparent in the very definition of *recombinant DNA* — a term that derives from the actual physical recombination of previously unconnected, often evolutionarily distinct genetic molecules. We define recombinant DNA as DNA molecules derived from different sources that have been artificially spliced together in vitro — outside living organisms — to form novel hybrid DNA molecules not normally encountered in nature.

Because modern genetic engineering techniques based on recombinant DNA technology permit genetic exchanges between species that do not normally interbreed, they offer us the opportunity to transcend the limits imposed by nature on hereditary processes. Genes no longer have to be shepherded from one generation to the next, for example, in vaguely defined groups associated with a desired phenotypic trait. Using recombinant DNA techniques, scientists can manipulate genes individually — by directly modifying the DNA molecules in which genetic information is encoded. Thus, recombinant DNA technology has the potential to transform the genes of all species into a global resource that can be used to shape novel life forms obedient not to the dictates of natural selection but to those of the human imagination.

The Art of Gene Splicing

In 1973, two decades after Watson and Crick published their revelations about the architecture of the DNA double helix, a historic event took place that marked the beginning of modern genetic engineering. Herbert Boyer, a researcher at the University of California Health Science Center in San Francisco, and Stanley Cohen, at Stanford University, succeeded in ferrying a recombinant DNA molecule containing DNA sequences from a toad and a bacterium into a living bacterial cell. There, to almost everyone's astonishment, the foreign toad DNA was copied and biologically expressed in protein.

For the first time, scientists had choreographed genes from the cells of an evolutionarily advanced species to dance in the cells of a distantly related species. Their technique of artificially inserting bits of foreign DNA into a bacterial host so that the DNA sequence can be rapidly copied is called *gene cloning* (see Figure 5.1). Most of us think of cloning as the creation of genetically identical cells or organisms derived from a single ancestral cell. But geneticists use the same term to describe the production of identical copies of genes, in addition to copies of cells or entire organisms.

Gene cloning is the single most important process in recombinant DNA technology. For even in its most basic form, it reveals the power of recombinant techniques to break down gene-bearing DNA molecules into smaller, more manageable pieces that can be isolated, analyzed, copied or manipulated to mass-produce specific protein products. In essence, gene cloning requires four simple, but clever choreographed, steps: (1) cutting and joining genes to create recombinant DNA; (2) transporting recombinant DNA into host cells; (3) growing recombinant genes in host cells; (4) screening cells for the desired recombinant gene.

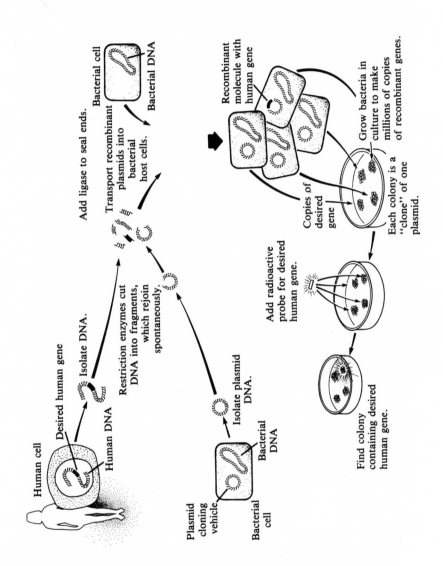

FIGURE 5.1: An Overview of the Gene Cloning Process

Step 1: Cutting and Joining Genes

The first step in cloning a gene is to shatter the long DNA threads in a cell of an organism harboring a desired DNA sequence into fragments of more manageable size (see Figure 5.1). There are a variety of ways to shatter a DNA molecule once it has been extracted from a living cell: vibrating it with ultrasonic waves, digesting it with chemicals, stirring it, like a milkshake, in a high-speed blender. But these are relatively crude techniques, cracking the molecule in unpredictable ways. What is needed is some sort of molecular scalpel — a precise method of chemically dissecting chromosomes.

Restriction Enzymes

Not until 1970, with the discovery of a new class of specialized enzymes in bacteria, did such analytical tools appear. These enzymes are called *restriction enzymes* because they tend to "restrict," or prevent, foreign DNA from another species from intruding on the harmonious genetic processes of bacterial cells. When foreign DNA — from an infective virus, for example — is injected into a bacterial cell, restriction enzymes swarm over it and chemically disable it. The enzymes accomplish this feat by first recognizing the DNA as alien, then binding to it and finally slashing the DNA molecules to ribbons. At the same time, the enzymes manage to leave the bacterial DNA completely unscathed.

At first, restriction enzymes seemed to be little more than an intriguing but minor bit of microbial natural history — a mechanism by which cells could distinguish and defend "self-DNA" from "none-self DNA." But, in time, it became clear that these enzymes are widespread among species of bacteria. More important, the mechanisms by which they are able to dispatch alien DNA turned out to be incredibly precise.

In 1973, the year Boyer and Cohen demonstrated their recombinant alchemy, geneticists had perhaps eight different restriction enzymes at their command. A decade later, scientists had identified roughly 300 different restriction enzymes from a variety of bacterial species. To distinguish one enzyme from the other, each was assigned a somewhat cryptic scientific name. Consider, for instance, the restriction enzyme *Eco*RI, a resident of one strain, designated R, of the ubiquitous human intestinal bacteria *Escherichia coli*. The first letter of the enzyme's name is taken from the first letter of the bacteria's first, or genus, name; the next two correspond to the first letters of its second, or species, name; the R identifies its designation in the lineage of the *E. coli* clan; and Roman numeral I reveals that this is the first of several known restriction enzymes in this organism. In a similar fashion, the *Pst*I enzyme turns out to be one of the restriction enzymes in a species of bacteria called *Providencia stuartii*.

Restriction enzymes tend to share certain structural features that enable them to search out and destroy foreign DNA by homing in on a specific target sequence — usually four to six base pairs in length — on a long, meandering DNA strand. The *Eco*RI enzyme, for instance, has a built-in ability to find the base sequence G-A-A-T-T-C (see Figure 5.2). Each restriction enzyme can then slice the offending strand of foreign DNA in two at a specific base-pair junction along the strand — predictably and with unerring accuracy. *Eco*RI, for example, always cuts each DNA strand precisely between neighboring guanine (G) and adenine (A) nucleotide bases lying side by side on the same strand.

Palindromes

Like nuclear physics — with its mirror image subatomic worlds of particles and antiparticles — molecular genetics sometimes offers glimpses into nature that can dazzle us with their sheer unadorned symmetry. Few of them can match the crystalline complementary design of the DNA double helix itself. But even minor genetic players like the diligent restriction enzyme can dazzle us with this same light.

Take, for instance, the simple fact that the stretches of DNA that most restriction enzymes are chemically programmed to recognize turn out to be symmetrical base sequences geneticists call *palindromes*. In linguistics, a palindrome (from the Greek,

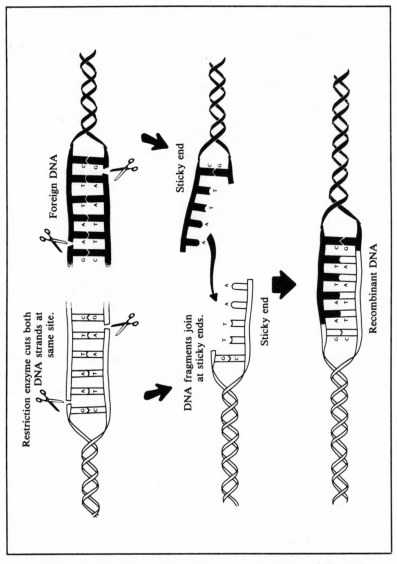

FIGURE 5.2: The Restriction Enzyme Activity of EcoRI

for "running back again") is defined as a word, or cluster of words, that reads the same forward or backward. The word *radar* is an English-language palindrome, as is the sentence "ABLE WAS I ERE I SAW ELBA." The date 1881 is a numerical palindrome. In molecular genetics, a palindrome is a chemical phenomenon that exhibits this same peculiar property. It can be defined as a base sequence in a DNA double helix that is the same when one strand is read left to right or the other is read right to left.

Each strand of a DNA double helix has opposite polarity. That is, the DNA sequence of each is read exclusively in one direction during the transcription of that sequence into messenger RNA, and that direction is exactly opposite to that of the complementary strand. The following DNA sequence is the specific palindrome recognized by the *Eco*RI restriction enzyme:

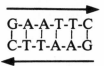

In this sequence, reading the upper strand from left to right results in precisely the same base sequence as reading the lower strand from right to left.

Why most restriction enzymes require such base-pair symmetry to shatter the double helix is unknown. Yet the consequences for recombinant DNA techniques are profound. Many restriction enzymes make incisions at the same point in each strand's palindromic sequence (see Figure 5.2). Because the identical target sequence is found on both strands of the double helix, the restriction enzyme creates two separate single-stranded breaks. Each cut is staggered — offset to the right or left of the midpoint of the palindrome sequence. As a result, each cut leaves a pair of single-stranded DNA "tails" that are not only identical in length and base sequence but also complementary.

The chemical characteristics of these two single-stranded tails are critically important to the prospective gene splicer. Because each tail is unpaired and chemically eager to bond to a complementary mate, it is referred to as a "sticky" end. Each tail is also complementary to the other. Our familiar *Eco*RI enzyme, for example, leaves behind one tail with an A-A-T-T base sequence and another tail with a T-T-A-A base sequence. If you recall the base-pairing rules for DNA — always an adenine with a thymine — you will notice that each strand is the identical four-base complement of the other.

Because each type of restriction enzyme can slice the double helix wherever it encounters a corresponding target site, it can slice the molecule into a multitude of DNA fragments. Depending on the size and complexity of the original DNA double helix, there may be millions of such fragments. Each enzymatic cut creates another pair of complementary single-stranded tails. Thus, half of all freshly cut ends have base sequences that are complementary mates of the other half. This means that the tail of each segment of DNA is potentially capable of bonding not just to its former mate but to *any* piece of DNA that has a complementary, single-stranded tail.

It is this random reunion of enzymatically shattered DNA that creates the new combinations of DNA sequences so vital to recombinant DNA technology. For after treatment with a restriction enzyme, DNA sequences from different species are as likely to rejoin as are pieces from the same species. In both cases, the sticky, single-stranded tails of each segment of DNA act as if they were chemically compelled to couple with the sticky tails of other nearby DNA segments. In fact, this stickiness is nothing more than the weak chemical affinity that always exists between complementary bases — precisely the same glue that holds every spiraling double-helix molecule in place.

A Recipe for Recombinant DNA

Imagine that you wanted to brew up a batch of recombinant DNA of your own — a blend of a bit of your own human DNA and that of foreign bacterial genes. In its simplest form, your recipe would proceed something like this.

You would begin by gently extracting glistening chromosomal threads of DNA both from a sample of cells in your body and from a pure laboratory colony of bacteria. Then you would combine the two batches of DNA in a single test tube, where you would let the DNA mix with a minute dollop of a single variety of restriction enzyme. Depending on which enzyme you chose, this process might shatter the 46 meandering double-helix molecules your cells possess into hundreds of thousand pieces, ranging from about 1,000 to 40,000 base pairs in length. At the same time, it would split the circular DNA double-helix molecule in each bacterial cell wherever the enzyme encountered its base-pair target. (In bacteria, DNA is almost always found in naked, unadorned rings of nucleic acid rather than in the complex, tightly coiled ropes of DNA and protein found in the chromosomes of humans and other higher organisms.)

At this point, you could sit back and watch recombinant DNA emerge on its own. In a matter of minutes, the millions of broken double helices would reassemble — one single-stranded sticky end to another — in a spontaneous chemical reaction. These reunions would be an entirely random affair. That is, the gradually lengthening DNA molecules would not simply be chemical re-incarnations of original chromosomes; they would be genetic mosaics — unpredictable hybrid molecules harboring a combination of your genes and bacterial genes.

Finally, you would add a small quantity of another enzyme ingredient, *DNA ligase*, to add the finishing touches to your recombinant DNA brew. This enzyme seals any exposed gaps in the rebuilt backbones of the new double-helix molecules, cementing the rejoined fragments securely into place and removing any lingering chemical scars that might remain.

The molecules that remain can legitimately be called recombinant DNA. In a sense, they are genetic chimeras — the unnatural progeny of human and bacterial DNA. In Greek mythology, the chimera was a fire-breathing she-monster, a fantastic, improbable blend of three different species — lion, serpent and goat. The imaginary DNA chimeras we have just cooked up bear scant resemblance to their fierce namesake; they look more like moist, microscopic noodles than mythological monsters. But like the chimera of myth, they represent a fusion of species not normally countenanced by nature. More important, the same basic formula that bred them can, in principle, be used to create other novel genetic chimeras from virtually any combination of species on earth.

Step 2: Transporting Recombinant DNA

We have now concocted recombinant DNA — DNA molecules that are genetic mosaics, or chimeras, of two or more species. But it is not enough to simply splice together genes from different species into laboratory glassware. Recombinant genes will lie idle until they have been transported inside a living host cell, where they can be transformed into functional RNA molecules and proteins. The second stage in the cloning of recombinant genes, then, is to move these hybrid DNA molecules from test tube to cell (see Figure 5.1).

Shuttle Systems for Genes

The two most important natural cloning vehicles — or *vectors* — for inserting foreign, passenger genes into a target cell are viruses and *plasmids*. A virus is little more than a bundle of genes wrapped in a protective protein package for delivery. Existing on the biological boundary between the living and the nonliving, a virus is incapable of carrying out its own replication or metabolism except by injecting its nucleic acids into the interior of a cell and genetically parasitizing it. Since the early 1960s, geneticists studying *phages* — viruses that prey on bacteria — have known that viruses are capable of carrying bits of bacterial DNA during their promiscuous passage from one host to the next. It is this natural gene-shuttling ability that has earned viruses a place as routine players in recombinant DNA experiments.

Plasmids, unlike viruses, are native residents of cells. Tiny self-replicating circles of DNA, they were discovered drifting in the cytoplasm of bacterial cells (see Figure 5.1). They are small compared with a cell's central gene-bearing loop of DNA, measuring on the order of hundreds of base pairs, rather than millions, in length, and their genes are not usually essential to normal bacterial metabolism. Thus, plasmids can readily be passed from one cell to the next — independent of the mainstream passage of genes during cell division — without disastrous consequences. However, plasmids are also capable of fusing with the main chromosomal loop of bacterial DNA and, after breaking free again, tearing away a few neighboring mainstream genes. This means that during the course of their nomadic journeys, plasmids, like viruses, can transport genes naturally throughout a population of suitable bacterial hosts.

Over the years, geneticists have learned how to modify these and other natural vectors into even more efficient carriers of recombinant DNA. They have succeeded, for example, in genetically disarming virulent viruses, accelerating the replication rates of some plasmids and expanding the gene-carrying capacity of both viruses and plasmids to make room for a larger load of recombinant DNA.

But recombinant genes are more than passive cargo. To function inside a foreign cell, they must be equipped with special control sequences of DNA — promoter sequences to which enzymes can attach during transcription of DNA to RNA, for instance — to ensure faithful expression of genetic messages once they have been delivered. This regulatory information represents another crucial element in the design of a customized vector.

Today there exists a long, and growing, list of specialized viral and plasmid vectors — proven recombinant DNA shuttle systems designed for a variety of jobs. Because each vector has its own replication rate, base-pair capacity and range of host species it can penetrate, together they offer a wide range of options for transporting genes. And as each new vector is developed, it is added to the list — creating a sort of mail-order catalog of "used vehicles" that can be shared by members of the scientific community all around the world.

Loading a Gene on the Vector

By exploiting natural gene vectors, genetic engineers have found a way to mobilize recombinant DNA. The trick is simply to join the DNA sequence to be transported to an appropriately modified plasmid or virus (see Figure 5.1). If, for example, the genes of choice were human genes, we might, as in our previous imaginary recombinant DNA experiment, simply douse both naked human DNA and plasmid (or viral) DNA with a quantity of one type of restriction enzyme. If a vector has been chosen wisely and harbors at least one of the enzyme's target base sequences, this procedure should slice DNA from both species into a tossed salad of truncated DNA molecules — each trailing a short wisp of one-stranded DNA that is complementary to others. This multitude of DNA fragments gradually begins to congeal into longer, double-stranded segments of DNA.

Again, this chemical reaction occurs spontaneously and without a controlled pattern to the multitude of tail-to-tail matings. But if our DNA sample is large enough, chance alone dictates that somewhere in this simmering nucleic acid soup will drift at least a few recombinant DNA molecules that consist of a reconstituted plasmid loop into which the human gene we wish to clone has been integrated. Thus, the desired gene has been effectively isolated from its fellow human genes and, mounted on a plasmid vector, is now ready to journey into a bacterial host cell.

Step 3: Growing Recombinant Genes in New Cells

In multicellular creatures like ourselves, it is not easy to isolate a single desirable gene. The cells of the human body harbor tumultuous populations of interacting genes. Some genes are programmed to switch off and on at precise times during human growth and development. Some fade into quiescence in specialized cells dedicated to the manufacture of only a relatively few gene products. Others either cease to replicate or else do so at a ponderously slow pace.

Gene cloning is one way to sidestep such genetic complexity. By cloning a human gene in a bacterial cell, the genetic engineer not only physically separates a specific DNA sequence from all the rest but also creates a system that can churn out identical copies of that gene with all of the speed and efficiency with which bacteria and other microorganisms routinely copy their genes (see Figure 5.1).

Bacteria divide very rapidly. Given adequate nutrition and space, a single bacterial cell, undergoing cell division every 20 minutes, could produce offspring of a mass greater than the mass of our planet in less than two days. If a human recombinant gene is stitched into a bacterial chromosomal loop, it will be replicated each time a bacterial cell divides in two. If the human DNA sequence is spliced into a more rapidly replicating plasmid vector instead of the bacterial DNA ring, the pace at which the human gene is duplicated can be accelerated hundreds of times.

This means that simply by riding on a plasmid, a single copy of one of your genes could quickly be cloned into trillions of exact replicas. Furthermore, if that recombinant gene were first properly outfitted with a promoter site and other DNA control sequences, it might not only be copied but also interact harmoniously with the metabolic machinery of the bacterial host. If the cloned gene were expressed, the bacterial cell would have been transformed into a miniature, genetically engineered factory capable of spewing out a potent human hormone or other gene-coded product in quantities that would not normally be tolerated inside a living human cell.

To minimize surprises, scientists have generally chosen to clone recombinant DNA in cells of familiar host species. No microbe has been more exhaustively studied than *Escherichia coli*, or *E. coli*, a species of cigar-shaped bacteria that routinely resides in the human digestive tract. To lessen the chance of genetic exchange between wild populations of intestinal bacteria and experimentally modified recombinant ones, scientists have deliberately crippled certain strains of *E. coli*. Addicted to laboratory nutrients, supersensitive to soap and stripped of genes that once activated their own DNA-swapping plasmids, they have been bred to favor life in test tubes over life amid the lush microbial flora of the human intestine. In a sense, such bacteria represent a new breed of domestic animal — reared for the sole purpose of reliably expressing foreign genes.

The precise portal of entry for recombinant genes can vary according to the peculiarities of the host species in which genes are cloned. Both plasmids and viruses, for example, can be harnessed to transfer bits of recombinant DNA into *E. coli* and other alternative bacterial hosts. And mammalian cells can often be invaded if they are infected with a genetically modified viral pathogen. But other host species, such as yeasts — the ancient microbial allies of beer brewers and bakers of bread — often require special treatment. In order for plasmid-bound recombinant genes to get past the barriers of tough outer cell walls, cells have to be chemically removed with enzymes. Later, after recombinant genes have seeped through a thin underlying membrane toward a nucleus filled with chromosomal DNA, the protective coats will be artificially replaced.

Yet regardless of what restriction enzyme, vector or host species we select, one daunting task still faces us. We must pinpoint the exact location of the pure colony of host cells carrying a copy of the gene that we seek. For it is that cluster of cells that will eventually provide us with unlimited copies, or clones, of the gene.

Step 4: Finding the Right Recombinant Gene
Unfortunately, host cells make no distinction between recombinant genes that humans want to clone and those they don't. Any gene introduced into a cell is capable of being faithfully reproduced. Thus, if a broth of bacteria is seasoned with randomly recombined DNA from human chromosomes and plasmids, infected bacteria can be expected to display every imaginable mix of the two kinds of genes. The challenge then becomes to isolate a rare cell that contains the desired recombinant gene from the millions of other cells that do not.

Narrowing Down the Choices
One way to begin this crucial selection process is to separate bacteria with vectors from those lacking vectors by using genes

that confer bacterial resistance to a particular antibiotic drug to screen recombinant cells from nonrecombinant ones. First, bacterial genes that code for resistance to an antibiotic such as tetracycline are spliced into the DNA of the plasmid vector, along with its load of foreign recombinant genes. Next, the plasmids are introduced into bacteria that are normally killed by exposure to tetracycline. Finally, these genetically modified bacterial host cells are grown in glass dishes containing a nutrient jelly that has been laced with lethal levels of the antibiotic. Any cells that lack genes coding for resistance to the poisonous drug quickly perish. But those cells equipped with plasmids harboring recombinant genes resistance to the antibiotic will continue to thrive on the toxic medium. Each genetically modified survivor flourishes where it lies in the dish — creating a genetically uniform colony of cells. Each pure, geographically distinct colony contains millions of identical copies of the particular sequence of foreign DNA ferried by the plasmid that invaded it.

Tracking Down the Recombinant Gene
Now that only colonies containing recombinant DNA remain, the scores of different cloned genes must be systematically screened to find the still-hidden recombinant gene we want. Scientists have developed all sorts of clever schemes to pinpoint a desired DNA sequence — some of them extremely sophisticated. We will briefly explore a few of the basic methods for isolating a desired recombinant gene.

1. *Using an RNA Probe.* If the target gene is well known, one can purify a sample of messenger RNA molecules transcribed from the gene in the laboratory and attach a radioactive substance to it, thereby constructing what is known as an RNA *probe.* Because the base sequence of the labeled RNA molecule corresponds to that of the desired recombinant gene, the RNA probe will home in on and bond snugly to the target DNA sequences in the bacterial colony in which it occurs.

Sensitive photographic film is used to record the precise geographic location of those colonies harboring the radioactively labeled recombinant genes.

2. *Using Reverse Transcriptase or Antibody Probes.* Since cells possess a variety of different messenger RNA molecules — reflecting the multitude of genes from which they are transcribed — it is not always easy to purify a sufficient quantity of a particular type of messenger RNA to serve as a probe for a recombinant gene. In this situation, a remarkable enzyme called reverse transcriptase — found in viruses that employ RNA rather than DNA as a primary genetic molecule — can be used to synthesize fresh, single-stranded DNA replicas of the rare messenger RNA molecule. Once these complementary DNA copies are radioactively labeled, they can be released as DNA probes that home in on the target recombinant gene just like their RNA probe counterparts. Alternatively, probes can be fashioned from *antibodies* — special molecules produced by the immune system to find and fend off foreign substances called *antigens*. If the protein product of a target recombinant gene is injected into a laboratory rat, for example, it acts as an antigen, stimulating the synthesis of antibodies that can bind to and neutralize it. If these specific antibodies are fitted with radioactive labels, they can reveal the location of the desired recombinant DNA sequence by binding to the antigenic protein product as it seeps out of bacterial colonies harboring the target gene.

3. *Using the Genetic Code.* One of the most powerful strategies for tracking down recombinant genes is to use the genetic code itself. All one has to do is isolate a particular protein product inside a cell and determine the precise order of amino acid subunits in it. Then, with a table containing the genetic code in hand, one can figure out the corresponding sequence of bases that a DNA molecule would require to code for that chain of amino acids. This inferred DNA sequence can then be used as a recipe to cook up an artificial copy of the unknown gene in the laboratory. The synthetic gene can then

be labeled and released as another kind of radioactive DNA probe to track down bacterial colonies harboring recombinant copies of the same gene.

There are other ways to distinguish a desired recombinant gene from all the rest. But regardless of which technique is used, the goal remains essentially the same: to pinpoint the precise location of bacterial colonies containing the cloned genes. For once these genetically engineered cells have been located, they can be removed from the dish and grown in the laboratory as a source of endless replicas of the original gene.

Putting Recombinant DNA to Work
Once a gene has been cloned, it is available in quantity for further study. Cloning permits the mass production of specific DNA sequences. It transforms them from rare and elusive segments of DNA into pure and plentiful ones.

Constructing a Gene Library
One of the first tasks in exploring the genome, or genetic endowment, of any species is to assemble a large enough collection of recombinant bacterial colonies to contain cloned fragments of every DNA sequence in that genome. Geneticists refer to such a complete inventory of recombinant organisms as a "gene library" because it represents a living repository of a species' complete genetic makeup.

Building a gene library is not as difficult as it might seem. The process is simply a variation on the general gene-cloning scheme we have just described. Imagine, for instance, that you wanted to clone all of the genes in a single human cell — an estimated three billion nucleotide base pairs — so that you could begin to study each gene individually. To create this library of human genes, you would follow these basic steps (see Figure 5.3):

1. Remove chromosomal DNA from a sample of human body cells.

2. Treat this DNA with one or more restriction enzymes that can crack 46 double helices — each millions of base pairs in length — into fragments thousands of bases long.
3. Expose a virus vector to the same restriction enzymes, creating a second pool of DNA fragments with sticky ends that will bond with those in the first pool.
4. Mix the two batches of fragmented DNA together so that pieces of human and viral DNA link together in random combinations, allowing every gene in the human genome to become cargo for at least one virus.
5. Grow the recombinant viruses on glass dishes covered with a lawn of bacterial cells. As each virus infects a cell, it generates a horde of identical viral clones. With each replication cycle of these viruses, the human DNA sequence is duplicated along with the virus's own genes.

This set of viral colonies constitutes a human gene library. Each recombinant virus represents one "book" — one cloned human DNA sequence — in that library. Within the total base-pair "text" of all of those viruses lies a replica of every DNA sequence found in the 46 chromosomes of the original human cell sample. But here our library metaphor fails us. For while this population of recombinant viruses has effectively partitioned the genetic content of a human cell into discrete "volumes" of coded text, it has not indexed it for easy access. There is no card catalog in this library of genes. Building one is a tedious process of tracking down each gene in the library and identifying its precise DNA sequence and location. In the human gene library, this indexing task has scarcely begun. But, with the help of new laboratory techniques, the prospect of deciphering the entire human genome no longer seems impossible.

Making a Preliminary Map
Once a gene has been cloned, its more precious secret — the ordered sequence of base pairs — can also be determined. One way to begin to decipher a cloned DNA molecule is to use restric-

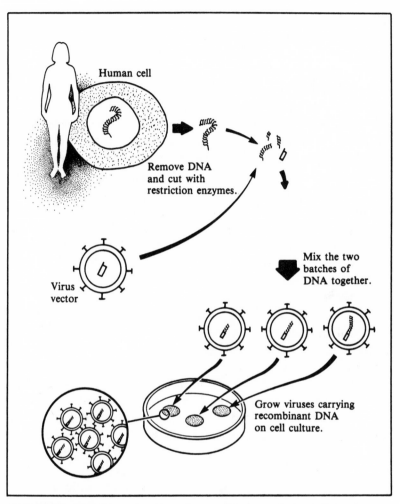

Human cell

Remove DNA
and cut with
restriction enzymes.

Virus
vector

Mix the two
batches of
DNA together.

Grow viruses carrying
recombinant DNA
on cell culture.

FIGURE 5.3: Steps in Creating a Human Gene Library

tion enzymes to create what is called a restriction enzyme map. In its simplest form, it involves treating an unknown stretch of DNA with a series of restriction enzymes. Each reaction, by generating its own distinctive set of DNA splinters, provides a few more chemical clues to the location of restriction enzyme target sites in the original molecule. Based on the length and number of fragments in each experiment, details about the geography of enzyme targets emerge. Like intersecting words in a crossword puzzle, data derived from each restriction enzyme bath can be combined to create a crude genetic map showing known base-pair target sites along the length of a DNA molecule. But a number of far more sophisticated techniques for reading the base-pair text of a gene in its entirety have become widely available. Although a gene must be cloned in large quantities before these rapid *DNA sequencing* methods are applied, they are not themselves recombinant DNA technologies. Instead, rapid DNA sequencing represents a powerful parallel technology that has developed concurrently with cloning and other recombinant strategies. It relies on an assortment of clever chemical tricks to expose each successive base pair in a piece of DNA, like frames of film in a camera.

Sequencing a Gene

One popular method, for instance, decodes DNA by allowing copies of an undeciphered strand of DNA to artificially replicate in four separate test tubes (see Figure 5.4). Then they are suddenly forced to stop. Each test tube contains all of the enzymes and nucleotide raw materials needed for DNA synthesis. But a small quantity of "counterfeit," radioactively labeled nucleotide component parts is added to each test tube. Altered adenine is added to one, thymine to another, cytosine to a third and guanine to a fourth. These nucleotides have been chemically modified to act as stop signals during the step-by-step synthesis of a DNA strand. That is, once one of these marked nucleotides has been enzymatically stitched into place in a growing DNA chain, it is incapable of bonding to the next nucleotide subunit that is added to the chain. As a result, DNA synthesis instantly grinds to a halt.

Like a child suddenly left chairless in a game of musical chairs, a radioactive terminal nucleotide in the suddenly quiet DNA strand in the first test tube mutely reveals its exact position in the chain. Because there are many DNA chains at different stages of completion replicating in this test tube, the abrupt chemical

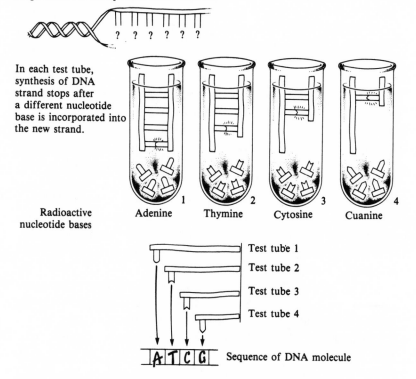

Single strand of unsequenced DNA molecule

? ? ? ? ? ?

In each test tube, synthesis of DNA strand stops after a different nucleotide base is incorporated into the new strand.

Radioactive nucleotide bases

1   2   3   4

Adenine   Thymine   Cytosine   Cuanine

Test tube 1
Test tube 2
Test tube 3
Test tube 4

A T C G   Sequence of DNA molecule

FIGURE 5.4: DNA Sequencing Technique

paralysis triggered by uptake of labeled adenine terminates at least a few DNA strands at every site in the sequence where adenine normally occurs. In the end, the first test tube contains an assortment of partially assembled DNA chains — all frozen in time, differing in length and terminating in a radioactive adenine base.

In a corresponding way, DNA molecules undergoing synthesis in the second test tube, spiked with chemically impotent cytosine, generate a different array of partially synthesized DNA strands. Again, the strands of DNA — this time capped by radioactive cytosines rather than adenines — vary in length. A parallel process takes place in each of the last two test tubes — one laced

with bogus thymine, the other with bogus guanine — furnishing
the rest of the required clues.

Molecular geneticists have instruments that allow them to
measure differences as small as a single nucleotide base in the
length of DNA strands. This ability enables them to determine
the precise position in the gene's linear DNA sequence of a
radioactively labeled adenine, thymine, cytosine or guanine base.
In this way, they can piece together the linear order of bases in
the DNA molecule — as if filling in blanks in a crossword puzzle.
Once the site of every base in the DNA molecule strand has been
solved in this manner, the gene can be said to have been
sequenced.

As a result of computerization and other improvements in DNA
sequencing techniques, pieces of DNA several hundred bases long
can now be routinely sequenced in a matter of only a few days.
Once this information has been gleaned from a stretch of DNA
the gene stands naked. Its message, so elusive in its natural state,
can be reduced to simple unadorned rows of typewriter characters
— rows of *A*'s, *T*'s, *C*'s and *G*'s depicting the four nucleotide
bases. These data can be stored in powerful computers, which
are programmed to scan them for signs of control signals, compute the amino acid sequences of their products or even compare them with known DNA sequences from other species.

Synthesizing a Gene Step by Step

After a gene has been decoded, it is possible to go one step further and use its base sequence as a recipe for cooking up an artificial copy of the same gene. Synthetic genes remain somewhat
of a novelty. For the moment, the process is slow, cumbersome
and limited largely to the assembly of structurally simple genes.

One way to make a gene is based on the deliberate interference
in DNA synthesis, in much the same manner already described
for rapid DNA sequencing. If adenine and thymine, two of the
ingredients for DNA synthesis, are placed in a test tube with the
requisite enzymes, long polynucleotide chains built of randomly
ordered A and T molecules will begin to form. This reaction is

halted in order to harvest a small number of A-T molecules, before longer polynucleotide chains such as A-T-T or A-T-A have time to form. If these A-T molecules are subsequently placed in a test tube of pure cytosine nucleotides, cytosines will be randomly added to the stubby chains. But, again, one can quickly retrieve some of these the instant they form base triplets — A-T-C molecules — leaving behind longer polynucleotides such as A-T-C-C. By repeating this process of incrementally adding one base to the growing DNA chain, it is possible to gradually synthesize a DNA molecule according to the base-pair recipe of one's choosing.

It is a tedious and frustratingly inefficient procedure. Because only a minute fraction of the polynucleotides at each step in the recipe have the desired base sequence, each successive harvest of molecules grows proportionally smaller — quickly dwindling to almost nothing. For this reason, the construction of long, complex DNA sequences, thousands of bases long, is not yet possible.

Synthesizing a Gene from Prefabricated Parts

An alternative technique, pioneered by H. Gobind Khorana during the mid-1960s, sidesteps this laborious base-by-base assembly of an artificial gene by synthesizing it instead from larger, prefabricated parts. Consider, for example, three DNA fragments that have already been synthesized using the previous step-by-step technique: a strand with the base sequence G-T-G-G-A-C-G-A-G-T, another with C-C-A-C-C and a third with T-G-C-T-C-A-G-G-C-C. If the first two strands are mixed together in a broth, their short, complementary stretches of bases mate, leaving an unfinished double-helix molecule that can be diagrammed like this:

G-G-T-G-G-A-C-G-A-G-T
| | | | | |
C-C-A-C-C

Its dangling string of six unpaired bases, on the right, is designed to neatly juxtapose with a complementary portion of the next polynucleotide fragment added to the brew. This union quickly zips together another six-base section of double helix. At the same time, it leaves still another single-stranded stretch to overlap, in turn, with the next polynucleotide unit added:

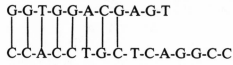

Even with such chemical shortcuts, synthesizing artificial genes is painstakingly slow, sometimes requiring days to piece together a few dozen nucleotides that a living cell could join in a matter of seconds. Nonetheless, it represents a milestone in genetic engineering. It has already permitted, for example, the construction of a human gene that is 514 base pairs long and codes for the antiviral protein called interferon alpha, which functions in bacterial hosts. Moreover, gene synthesis can provide a valuable tool for obtaining pure samples of genes that turn out to be especially difficult to isolate using more conventional means. As we described earlier in our discussion of DNA probes, it is sometimes possible to predict the recipe for an unfamiliar gene, knowing nothing more than the genetic code and base sequence of its messenger RNA transcript or the amino acid sequence of its protein product. The recipe can then be used to cook up the gene from scratch, without anyone's ever having seen the DNA molecule itself. Finally, the repetitious nature of laboratory gene manufacture has rendered the process highly suitable for computerized control. Already we are seeing early generations of programmable "gene machines," capable of stitching together a dozen or more polynucleotides in a night, following the prescriptions of geneticists as they sleep.

The Power of Recombinant DNA Technology

Already recombinant DNA technology, along with rapid DNA sequencing and other related laboratory techniques, has begun to offer profound new insights into the workings of genes. First, it has helped to establish the virtual universality of both the genetic code and the graceful molecular dances of genes in nature. Second, it has revolutionized the study of genes of multicellular organisms, including humans, that have long been inaccessible. In so doing, it has led to dramatic discoveries concerning the structure and function of genes in complex plants and animals. One such discovery is that their coded instructions for assembling amino acids into proteins — called *exons* — are often interspersed with long intervening segments of DNA — called introns — of largely unknown function, which are neatly edited from nuclear RNA molecules prior to protein synthesis. A second discovery is that the comparison of DNA sequences in different species can often serve as a faint chronological record of otherwise hidden evolutionary relationships. Third, recombinant DNA technology has already led to the complete chemical characterization of a number of relatively small viral genomes and has initiated that process in more complex species, including our own. Finally, by facilitating the exchange of genetic components between cells of different species, it has exploded the boundaries of comparative genetics. We are now just beginning to grasp the evolutionary ingenuity with which the genes of different species — however interchangeable their codes — are variously packaged, punctuated and controlled.

But the rewards from recombinant DNA technology will be practical as well as intellectual. Genetic engineering holds great promise in agriculture — for modifying the productivity, growth requirements and nutrition of food crops. In medicine, it is opening up new vistas in the design of vaccines, in the diagnosis of disease and in genetic counseling, and even the possibility of new genetic therapies for ancient hereditary diseases. In the biotechnology industry, it already has established its value in the manufacture of antibiotics, hormones and an assortment of other biologically active substances.

The possibilities seem positively endless. Recombinant DNA technologies have transformed us from passive observers of the global hereditary scene to active choreographers of genes — for motives that could range from simple scientific curiosity or altruism to profiteering or political demagoguery. So powerful is this new genetic art of manipulating genes that it already threatens to inundate us with more data about the structure and function of genes than we can possibly digest. Overwhelmed with this detail, we now sometimes find it difficult to even know what questions to ask of living biological systems — let alone what sort of impact new genetic technologies might one day have on hereditary processes that we have barely begun to understand.

For the first time, we find that we can command individual genes to do our bidding. In the future, we will forever be faced with the temptation to harness our knowledge of hereditary processes long before we have resolved the possible long-term consequences of our applications. If history is any guide, we can expect no shortage of ingenious schemes to reap rewards from our genetic engineering skills. But each of us has a responsibility to ensure that equal imagination is devoted to the search for ways to apply this knowledge wisely and to share its fruits with all humankind.

CHAPTER 6

BLAMING CRIME ON CHROMOSOMES: THE MYSTERY OF THE MAN WITH TOO MANY YS

GENETHIC PRINCIPLE
The vast majority of human hereditary differences are polygenic, or involve the interplay of many genes; therefore, it is a dangerous simplification to proclaim a causal relationship between human behaviors and so-called "defects" in human DNA.

Years later, Dante was dying in Ravenna, as unjustified and as lonely as any other man. In a dream, God declared to him the secret purpose of his life and work; Dante, in wonderment, knew at last who and what he was and blessed the bitterness of his life. Tradition relates that, upon waking, he felt that he had received and lost an infinite thing, something he would not be able to recuperate or even glimpse, for the machinery of the world is much too complex for the simplicity of men.

— Jorge Luis Borges, in "Inferno I, 32"

141

In 1961, a tall, healthy, normal-looking middle-aged man — with a history, it seems, of barroom brawling, especially during his youth — underwent a series of medical tests to determine why he had recently fathered a genetically abnormal child. The tests did not identify the cause of the baby's illness. But they did inadvertently reveal something that astonished the man's doctors. Every somatic cell in his body had one too many chromosomes — 47 — and in a combination, or genotype, that had never before been reported in the literature of medicine.

In the shorthand of the medical geneticist, the man's genotype was 47,XYY. In addition to the usual 46 chromosomes, including 44 *autosomes* (22 pairs), or chromosomes that are not sex chromosomes, and the X and Y sex chromosomes normally found in a human male, he had an extra Y chromosome.

The report of the world's first recorded case of an XYY man was published that year in the British medical journal *Lancet*, making minor medical history. Today we know that the XYY genotype is a relatively rare chromosomal condition, occurring in about one in a thousand newborn male infants. And we know that it can be traced to a failure in the distribution of chromosomes, called *nondisjunction*, that takes place in paternal reproductive cells during meiotic metaphase.

During act 2 of meiosis, as you may recall, duplicate pairs of chromosomes normally separate and drift toward opposite poles of the dividing cell (see Figure 6.1). Occasionally, though, sex chromosomes fail to part and become trapped together, like Siamese twins, in one of the daughter cells; they may be two duplicate X chromosomes or two duplicate Y chromosomes. If the tandem structure turns out to be two Y chromosomes, the cell will eventually form a maverick YY sperm. When this sperm later goes on to fertilize an X-bearing egg from a normal female, the result will be a male embryo that has an XYY genotype. As the genetically abnormal embryo matures in its mother's uterus, it transmits its microscopic chromosomal error through mitosis to millions of other body cells.

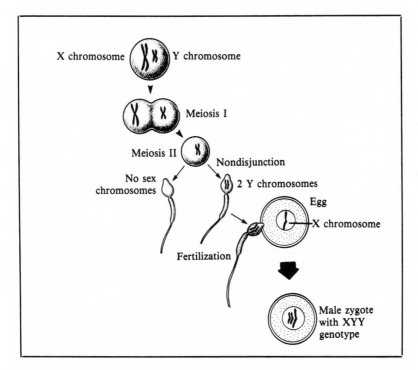

Egg

Fertilization

FIGURE 6.1: Meiotic Nondisjunction Leading to XYY Male

Karyotyping

The discovery of the XYY man was made possible by karyotyping, a revolutionary new laboratory technique developed during the 1950s that provided a direct microscopic inventory of human chromosomes. Before karyotyping, geneticists lacked any reliable method of detecting even gross chromosomal abnormalities such as the XYY condition. In fact, they did not even know for certain precisely how many chromosomes human body cells harbored.

In 1956, two researchers, Joe Hin Tijo and Albert Levan, made microscopic smears of lung cells from aborted fetuses that established with certainty for the first time that the human diploid number of chromosomes was 46. Their technique was quickly refined and adapted into an effective technique for assessing the chromosomes of adult patients.

The procedure was relatively simple. Human cells, from laboratory tissue cultures or freshly drawn blood, were treated with chemicals to halt the movement of chromosomes during mid-mitosis. Cells were then squashed, smeared across a glass microscope slide and stained for visibility. The entire quota of chromosomes in one cell could then be photographed through a microscope. The final step was to clip each chromosomal image from the print, sort the images according to size into rows of matched homologous pairs and paste them into a composite photograph known as a karyotype. Under such scrutiny, each pair of chromosomes in the human genome began to emerge with its own distinctive identity — based on its size, the location of its centromere, horizontal banding patterns and other traits. Suddenly, by comparing karyotypes of patients with inherited disorders to karyotypes of normal, healthy individuals, geneticists could begin to look for subtle structural differences in the human genome that they might correlate with a patient's disorder or disease.

Armed with this powerful new technique, research teams quickly set out to produce karyotypes of victims of a number of perplexing, seemingly inherited diseases in the hope of uncovering corresponding chromosomal defects. They eagerly set out on this search — without a clear grasp of any causal relationships between chromosome abnormalities and human disease. But their eagerness was perfectly understandable, for it reflected a genuine hope that they might find the first genetic clues to a number of baffling diseases that had been responsible for centuries of human suffering. And their search would win the researchers immediate rewards.

In 1959 alone, three previously enigmatic diseases were linked with chromosomal defects visible in karyotypes. People with Down's syndrome, a congenital condition that could leave children mentally retarded and weakened by a constellation of physical problems, were found to possess an extra autosome. They all had a triplet set, or *trisomy*, of chromosome pair number 21; other chromosomes were normal. Patients with Klinefelter's syndrome, a rare disease that tends to feminize breast development and other secondary sex characteristics in men and render them sterile, turned out to have an XXY genotype. And women afflicted with Turner's syndrome, a condition that causes infertility and immature sexual development, proved to have XO genotypes, lacking the second X chromosome typical of females. All of these conditions would eventually be traced, like the XYY disorder, to errors in the orderly partitioning of chromosomal pairs during meiosis (see Figure 6.2).

These were intoxicating discoveries for medical geneticists and left them justly proud. And these discoveries were quickly followed by others. In this atmosphere of optimism, it is not difficult to imagine the flurry of excitement created by the brief clinical report in *Lancet,* just two years later, of an anonymous, slightly aggressive middle-aged man who had been born with a previously unknown sex chromosome defect known as XYY. In the same journal, less than a year later, the tantalizingly proposition was put forward that the same genotype might possibly be linked to a life of crime.

The Crime Connection

On December 25, 1965 — only four years after the initial XYY report — the prestigious British journal *Nature* published a short article that sparked more than a decade of fierce debate and controversy about the so-called XYY syndrome — both within the scientific community and beyond. Titled — somewhat provocatively, it would turn out — "Aggressive Behaviour, Mental Sub-normality and the XYY Male," it described a recent survey of the karyotypes of 197 men. All of them happened to be patients in Carstairs, a high-security mental hospital in Scotland. These men had been chosen specifically because their presence at Carstairs marked them as "mentally sub-normal male patients with dangerous, violent, or criminal propensities."

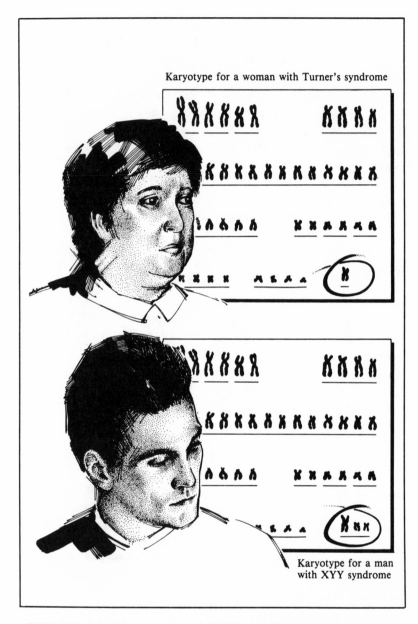

Karyotype for a woman with Turner's syndrome

Karyotype for a man
with XYY syndrome

FIGURE 6.2: Karyotypes of XYY and Turner's Syndromes

The article was written by Patricia Jacobs, a highly respected cytogeneticist who had worked in British laboratories under some of the leading pioneers in the field of sex chromosome abnormalities, and four colleagues. It offered some rather startling findings. In making karyotypes of cells in blood samples taken from patients during routine physical examinations, the researchers had discovered an unexpectedly high number of visible chromosomal abnormalities — a total of 12 in all. Seven men, about 3.5 percent, turned out to have the XYY genotype that was the target of the study. (The remaining five displayed an assortment of defects: one was XXYY; another had a patchwork, or mosaic, of XY and XXY cells; three showed signs of various autosomal errors.)

The XYY karyotype was still relatively new to science, and the frequency of XYY males in the general public, beyond the barred windows of the mental institution, remained unknown. But 3.5 percent struck Jacobs and her team, at least on theoretical grounds, as unusually high — perhaps 20 times higher than normal or more. Jacobs also noted, almost parenthetically, that her XYY subjects, averaging over 1.8 meters, or 6 feet, in height, seemed to be noticeably taller than genetically normal males. But without a doubt, the most damning factor in this XYY study was the uniform history of mental illness, aggression or both in its subjects. That history was officially documented; now so was the high incidence of XYY karyotypes. While the relationship between the two might remain uncertain, the implications seemed less so. "At present," concluded the authors, "it is not yet clear whether the increased frequency of XYY males found in this institution is related to aggressive behavior or to their mental deficiency or to a combination of these factors."

Despite the note of uncertainty in the statement, it nonetheless seemed to imply that the XYY genotype had already been linked to male aggression or mental defects. Earlier studies of related sex chromosome defects had already speculated on the possibility of such a relationship. But now, for the first time, the association between XYY males and abnormal behavior was indelibly etched in the pages of one of the world's most prestigious scientific journals. Could one extra Y chromosome somehow alter the genetic destiny of a newborn male child — predisposing him to a life clouded by mental illness, violence or crime? And if so, why? Moreover, what ought responsible, genetically "normal" citizens do to protect their communities from the possible dangers of these free-roaming hordes of "chromosomal defectives"?

The Role of the Y Chromosome

What was distinctive about the XYY genotype from the beginning was that unlike more severe conditions such as Down's syndrome and other more severe chromosomal defects, it was accompanied by no predictable set of physical or mental symptoms that could be used to reliably diagnose the condition. Down's syndrome, for instance, was not detected solely by the presence of an extra chromosome 21. People with Down's syndrome might display any of a constellation of defects — from distinctively slanted eyes and a deformed skull to mental deficiencies and congenital heart disorders. In contrast, XYY males would turn out to display few predictable characteristics. The most notable: increased height and lowered fertility when compared with genetically normal males. Most other stereotypical XYY symptoms would later be traced to anecdotal evidence, conjecture or statistical sleight of hand.

The first documented XYY case had not been diagnosed on the basis of bizarre physical or behavioral symptoms but, inadvertently, by medical fluke. Nor did Jacobs's influential but preliminary study pretend to establish a firm causal link between the XYY genotype and the tragic behavioral histories of the men at Carstairs. For no one had yet begun to dissect the complex question of what, if anything, genes had to do with those behaviors.

Part of the excitement over the XYY syndrome lay in the hope that it might eventually shed light on broader questions about the genetic basis of behavior in normal XY males. The line of reasoning went something like this: Men have a Y chromosome that women lack. Men tend to be more aggressive than women. If the Y chromosome were instrumental in altering sex-related bodily characteristics — from beard and biceps to the male sex organ — perhaps it was also essential for some of the behavioral patterns we consider masculine. And, if it were, perhaps the newly discovered XYY male — in a sense, a "supermale," with his double dose of the masculinizing Y chromosome — would serve as a natural test case to examine that hypothesis.

The logic of this hypothesis, it should be noted, highlights the importance of genetic factors in gender differences in human behaviors. There was, and continues to be, intense debate among respected biologists over the relative contribution of heredity and environment to the sex roles that males and females eventually assume in society.

We now know that most of the Y chromosome consists of *heterochromatin* — condensed regions of DNA containing genes that are inactivated and metabolically silent. Other stretches of DNA are complementary to those carried by its homologous partner, the X chromosome. Among the remaining relatively small number of genes unique to the Y chromosome are those that determine genetic maleness. During embryonic development, these genes activate the development of male reproductive organs, the production of sperm and the secretion of the masculinizing hormone testosterone. So the question might be refined to this: Could behavioral studies of XYY males, who presumably had twice the usual number of Y chromosome genes that confer genetic maleness, also illuminate the role of those genes in the behavior of normal males?

It was not a frivolous question. Earlier karyotype surveys of mentally retarded men in Swedish and British mental institutions had already alerted Jacobs to the possibility that an extra Y chromosome might predispose a man to aggressive behavior. In these studies of men with an extra X chromosome, those who also had a surplus Y chromosome (for example, XXYY genotype) were more likely to be found in high-security than in ordinary mental hospitals, compared with men having one Y chromosome (for example, XXY genotype). This finding had seemed to Jacobs to pose a question of almost mathematical clarity: Did one extra Y transform the multi-X genotype into a formula for aggression?

Nor is there any reason to believe that Jacobs's hypothesis was in any sense mean-spirited. Her stated aim was simply to pursue this line of reasoning to its logical conclusion. If the extra Y did somehow contribute to a more aggressive male personality in multi-X genotypes, the effect might be clearer in the simpler XYY genotype. Yet, whether well-meaning scientists were aware of it or not, the seeds for a bitter debate over XYY research were sown by the very nature of the Jacobs study.

First, the study focused on forms of human behavior — male violence and criminality — that represent a potential threat to society. Clearly, if karyotyping could somehow be used to quickly screen a population for early genetic indicators of antisocial behavior, it would be of enormous interest to police and other security forces. Such a program could also become a dangerous precedent leading to increased efforts to link other socially unacceptable behaviors — political terrorism, alcoholism, drug abuse, even homosexuality — however tenuously, to genetic markers that could be screened.

Second, the Jacobs study set out to establish a genetic basis for forms of behavior considered negative to most people. From the beginning, the XYY genotype was associated with disturbed, violent men in high-security insane asylums. Anyone found to possess the genotype tended to be stigmatized as violent or mentally ill — whether it was true or not. If the XYY abnormality had been investigated for a suspected association with more praiseworthy human behaviors — say, entrepreneurial talent, leadership skill or athletic prowess — any hasty hypothetical association between chromosome and character would have been less likely to cause personal harm. But once a young boy, for instance, was found to possess an XYY genotype, he could be marked for life — burdened with an unearned scarlet letter in the form of a second, or supernumerary, Y.

Finally, the XYY study proposed to apply what at the time

was one of the most powerful new tools in medical genetics — karyotyping — to one of the most politically powerless groups in our society — prisoners who were mentally ill. It could be argued that they were precisely the people who eventually might benefit most from such research, through new treatments or medical care. But Jacobs's XYY study, and ones that followed it, routinely neglected to include as subjects XYY males who were free and functioning normally in society. By focusing exclusively on inmate populations, these studies at least gave the appearance of searching for genetic confirmation that there was a biological basis for their subjects' powerlessness. At the same time, XYY researchers tended to ignore or downplay a host of less exotic environmental factors — ranging from family history to childhood nutritional deficiency — that could conceivably have contributed to their subjects' mental disturbances.

The Crime Connection Gains Credence

The same features that would later cause some scientists to openly question the value of XYY medical research made it an attractive science news story for the media. The possible link between heredity and crime has long been a topic of both scientific and public fascination. Here was a new opportunity to recycle that issue, in the light of the latest technological advances in genetics. Jacobs's 1965 survey was followed by a surge of similar studies that eagerly pursued the aggression hypothesis she had put forward in mental hospitals and penal institutions around the world. A number of them confirmed the disproportionately high rates of XYY males Jacobs had reported. But none succeeded in casting new light on any causal connection between chromosomes and criminality.

Nonetheless, these studies often received wide coverage in newspapers and magazines and on television. With disturbing frequency, the still highly preliminary XYY findings were publicly distorted and sensationalized. In the process, the media uncritically fostered an erroneous view of the supernumerary Y chromosome as the bearer of some sort of vaguely sinister "criminal genes." During this period, *Newsweek*, for example, published a story on XYY males titled "Congenital Criminals." More dramatically, in 1968 the XYY story made national headlines in the United States with a rash of reports claiming that Richard Speck, the notorious criminal who had been convicted of murdering eight student nurses in Chicago, had an XYY genotype.

The reports were untrue and were later retracted. But they fed a growing public perception that populations of XYY men not only were genetically "driven" to live lives of violence and crime but also might represent a hidden menace to society. Some scientists even began to propose genetic screening programs designed to diagnose the XYY abnormality prenatally. Since there existed no medical remedy for XYY fetuses, the implication was that informed parents might simply choose to abort their unborn XYY males before these offspring could mature to act out their presumably genetically preprogrammed "behavioral fate."

At first glance, the notion of systematically aborting babies bearing the XYY genotype must have struck some people as a eugenically sound plan — a pragmatic approach to gradually ridding the human gene pool of the newly defined XYY defect. If so, they were mistaken. Because the genetic missteps responsible for this trichromosomal state can occur anew in the meiotic cell divisions of each generation of parents, any such dreams of hereditary hygiene in the case of the XYY disorder would be illusory.

By 1968, defense attorneys in Australia and France had already tried to take advantage of this misconception by requesting reduced sentences for clients on trial for murder — simply by showing them to be XYY. During the late 1960s, a U.S. government agency called the Center for Studies of Crime and Delinquency, a branch of the National Institute of Mental Health, concluded that the "XYY syndrome" fell under its broad mandate in the areas of crime, delinquency and mental health. As a result, it began to fund projects that, for example, carried out mass XYY screenings of juvenile males held in detention and treatment

facilities in Maryland. And in 1969 it sponsored a major conference on XYY research, lending scientific legitimacy to the connection between XYY men and crime.

By the early 1970s, a more disturbing trend had developed.
Maternity hospitals in the United States, Canada, Denmark and
England had begun to permit mass screening of newborn infants
to see if they were XYY. On the surface, such chromosomal
testing, even for the enigmatic XYY defect, seems perfectly defen-
sible. After all, the same karyotype that might identify an XYY
baby could also reveal valuable information about more damaging
genetic diseases. And even if the XYY genotype could not yet
be conclusively linked to criminal aggression, it seemed that
parents of an XYY baby at least had the right to know that such
a possibility, however remote, existed.

The Controversy Continues

One study, carried out by Harvard child psychiatrist Stanley
Walzer and pediatrician Park Gerald, eventually became the focus
of a heated public debate about the XYY controversy. In 1968,
they had initiated an experimental genetic screening program at
the Boston Hospital for Women that would produce karyotypes
of male infants born in the maternity division within a day or
two of birth. The purpose of the program was not only to identify boys with XYY genotypes but also to track them systematically for a period of years. Through periodic home visits,
psychological tests and teacher questionnaires, the study would
record each child's behavioral development in an effort to identify abnormalities.

Because no medical treatment existed for the XYY condition
— no prescribed regimen of hormones, diet or drugs that might
alleviate any anticipated behavioral disorders — the researchers
could not hope to treat XYY children displaying any of the
expected signs of aggressive behavior. But they did propose to
offer families what they termed "anticipatory guidance" — professional counseling sessions to help them cope with periodic
behavioral problems their XYY boys might develop. In the end,

the hope was to amass enough information to be able to predict any XYY baby's risk of developing aggressive or sexual behavioral problems.

But the project would never achieve these ambitious aims. By 1974, a number of scientists and concerned private citizens had begun to challenge both the scientific validity and the ethical implications of the Boston XYY project, and they publicly called for its halt. The debate attracted considerable media attention and, in part because many scientists perceived it as a direct threat to their freedom to choose their own area of scientific research, quickly grew more rancorous. Among those leading the challenge were geneticists Jonathan Beckwith of Harvard and Jonathan King of MIT, along with a Boston political organization called Science for the People.

In 1974, in an article in the British journal *New Scientist* titled "The XYY Syndrome: A Dangerous Myth," Beckwith and King outlined their objections to XYY research. From the beginning, they contended, the majority of XYY studies had been fundamentally flawed. They had been poorly designed, filled with logical inconsistencies and crippled by inadequate comparisons with matched, normally functioning XYY males as controls. The studies had often been precariously built on simplistic definitions of extremely complex human behaviors — including aggression, social deviance and crime. And they had persistently downplayed the possible role of environmental factors in influencing traits too readily attributed to chromosomes.

But at the core of their critique was a concern for the serious ethical questions raised by a study such as Walzer's. First, researchers failed to obtain fully informed consent from participating parents. Because they were not told of the precise purpose of the karyotype tests or of the ambiguous medical findings from previous XYY studies, parents might not understand the dilemmas that could arise from an XYY diagnosis in their children. Second, if a child was diagnosed as XYY at birth and his family was informed of this diagnosis, that innocent child could be unfairly labeled at an early age as genetically prone to aggression and violence. This label could also contribute to a childhood setting in which a level of anger quite acceptable in a normal XY boy would be treated with undue concern by fearful parents aware of their child's genotype. In the end, this distortion could generate new behavioral problems. In other words, an XYY child could display abnormally aggressive behavior as a result of his role in the experiment — as the victim of a tragic self-fulfilling prophecy.

Beckwith filed a formal complaint with a Harvard research review committee, requesting that the body order a halt to the project on the grounds that it posed potential harm to its subjects and violated fundamental ethical principles of medical research. The committee eventually refused — but not before press conferences and public meetings brought the XYY issue into the public forum, earning additional support for the XYY critics. By 1975, faced with a growing public outcry over the ethical issues raised by his study, Walzer announced he would stop screening Boston newborns for the XYY genotype. Before long, scientists conducting a number of similar XYY studies, in Denver and elsewhere, decided to do the same — in part, to protect their patients from the public debate about their abnormality. Research on XYY males had simply become too hot to handle. And, for the most part, it has remained so to this day.

The climactic end of the Boston study closed a chapter in XYY research. But there remained many unanswered questions. Foremost was the mystery of the perplexingly high frequency — often 20 times higher than normal or more — of XYY males in high-security mental hospital settings. This mystery remains unsolved. At the same time, most XYY males — an estimated 96 percent — are thought to lead relatively ordinary lives, never seeing the inside of a prison or mental institution. Furthermore, medical experts — including Walzer and Gerald of the Boston XYY study — have now concluded that the list of supposed symptoms long thought to be associated with the XYY abnormality is so vague and inconsistent that the XYY genotype simply no longer merits the term "syndrome."

Tallness remains the most common physical characteristic found among XYY men. Some XYY males have facial skin problems, subtle speech and language difficulties or childhood reading disabilities (all traits, incidentally, that could conceivably lead to aggressive or antisocial behavior in a young boy). Other XYY males are reported to have suffered from slight hormone imbalances or repeated outbursts of anger. But to date none of these signs is known to recur with regularity. And in many cases they could have easily arisen from causes other than an extra Y chromosome. In the words of a 1979 medical review of the XYY condition: "There is no characteristic that is uniformly present in patients with 47,XYY chromosome constitution other than the fact that they do have an extra Y chromosome."

Lessons from the XYY Experience

If there is a single overriding lesson to be learned from the XYY debate, it goes far beyond the questions of the true clinical status, still largely unresolved, of XYY males, or even the unforeseen impact of XYY research on the lives of vulnerable human beings. Instead, this lesson must concern the incredible eagerness with which scientists and nonscientists alike have historically grasped at quick, socially convenient definitions of genetic disability or disease.

Overwhelmed by the complexities that underlie human differences, we are often too quick to judge one another on the basis of fragmentary genetic clues. In our impatience for easy answers to difficult questions, we run the risk of investing readily detectable hereditary traits with almost mystical prophetic powers. Like some prescientific shamanistic healer who finds people's fortunes revealed in fallen strands of hair or the discarded clippings of their fingernails, we unconsciously begin to extrapolate from the qualities of the part the qualities of the entire organism.

Regardless of which aspect of human genetic molecules we select for an early-warning system for human behavior — abnormalities in the number or structure of sex chromosomes, gene mutations or differences in DNA sequences so subtle they leave no visible phenotypic spoor — we risk being deceived. For most geneticists agree that the vast majority of human behavioral traits tend to be polygenic; they reflect the simultaneous interaction of a multitude of genes. And every element in our species' rich behavioral repertoire is unavoidably sculpted, at every stage of its development and action, by a myriad of environmental variables too numerous to possibly anticipate.

To further complicate matters, the manifestation of even a single human gene can at times prove virtually impossible to predict. A given gene may simultaneously affect a number of seemingly unrelated traits. Or it may be expressed, depending on its surroundings, in a spectrum of possible phenotypes — some of them indistinguishable from those coded by entirely separate genes. In the face of these and other genuine biological complexities in human inheritance, how can we expect to reliably read a person's behavioral destiny in the tea leaves of an isolated hereditary flaw?

The XYY story may serve as a valuable example of what happens when people insist on simple answers to problems in human behavioral genetics and apply these provisional answers in too great haste. But because the XYY story is an unfinished story, it can also leave us with a rather difficult personal dilemma. How are we supposed to react to the XYY genotype if it should, by chance, surface in our own individual lives? Should we be willing to welcome an unborn XYY child — however vague the suspected symptoms of his genotype — into the community of human beings?

Imagine, for instance, that you discover through prenatal tests that your unborn child is a boy and that, furthermore, he has an XYY genotype. Assuming your physician is knowledgeable about the XYY condition, he or she can only offer you uncertain counsel. Your son is likely to be born healthy and to lead a normal life, but some reports suggest that his extra Y chromosome may put him at slight statistical risk for developing subtle childhood behavioral or learning problems. The precise risk is unknown. The genetic mechanism is unknown. There is no reliable treatment. The choice of whether to proceed with the pregnancy is yours — the mother's and the father's. What choice would you make?

There are seldom easy answers or simple prescriptions for choices involving genetic defects and the life of an unborn child. Modern medicine has granted us increasing control over the size and the scheduling of our families. At the same time, it has placed the health of each planned child under closer medical scrutiny. Inevitably, parents are driven to try to use this new knowledge to ensure that they will have a healthy, "normal" child. But prenatal screening tests seldom yield results that can be used to predict the health of an unborn child with absolute certainty.

In the case of our hypothetical XYY fetus, at least, that ambiguity is so extreme that the presence of an extra Y chromosome can scarcely even be referred to as a genetic disease. The condition is not known to seriously harm human health and has a dubious and an unpredictable effect on the bearer of the surplus chromosome. Yet it is not unlikely that many parents of an XYY fetus who have previously been exposed to media accounts of the highly publicized XYY controversy might opt for abortion — if only because of the vague, poorly documented, but nonetheless disturbing, association of the XYY genotype with aggression and crime.

Considering the medical literature available to date, we would find it difficult to condone such a response. In this case, we find ourselves casting our hypothetical vote in favor of the chromosomally abnormal XYY child. (This is a strictly personal viewpoint; the decision, we believe, should be a matter of individual choice.) The reason: Every one of us is genetically "flawed" to some extent. Our genomes are likely to be laced with all sorts of errors — mutations, recessive disease-causing alleles, miscopied genes, chromosomal blunders — most of them, for one reason or another, masked yet fully capable of surfacing in a subsequent generation. Some scientists have suggested that these genetic "defects" can actually be favored by natural selection. Chromosomal duplications, for instance, are thought to have played a key role in the molecular evolution of genes coding for the hemoglobin molecule found in human red blood cells.

Since none of us can hope to claim genetic "perfection" — in the face of an unpredictably changing environment, it is, after all, an absurd notion — how can we pretend to expect it in a newborn child? At least until such time as a genetic abnormality has been indisputably shown to place severe limits on a child's future health, happiness or capacity to learn, it seems most wise to tolerate such imperfection, along with all of the rest. The XYY genotype has not been shown to place such limits on a child's future. Therefore, it cannot be said to have passed this simple ethical test.

One suspects that a certain elderly and anonymous man — the man with the dubious distinction of being the first XYY man on record — would, if still alive, wholeheartedly agree. So might literally hundreds of thousands of XYY males living relatively normal lives at every level of society around the world (though few of them today could possibly know of their chromosomal blemish). So too, one might argue with tongue only partially in cheek, ought every male on earth. For a compelling case could be made for declaring every genetically normal unborn male — equipped with a 46,XY genotype — at significant statistical risk for many of the same sinister characteristics that over the past quarter century have accumulated to stigmatize all 47,XYY males.

The reason: Compared with their Y-less female counterparts, men — throughout history and in virtually all nations — have shown themselves to be statistically far more prone to lives of aggression, violence and crime.

CHAPTER 7

GENETIC SCREENING IN THE WORKPLACE: PRIVACY AND THE HUMAN GENOME

> **GENETHIC PRINCIPLE**
> Information about an individual's genetic constitution ought to be used to inform his or her personal decisions rather than to impose them.

"We decant our babies as socialized human beings, as Alphas or Epsilons, as future sewage workers or future . . ." He was going to say future world controllers, but correcting himself, said "Future Directors of Hatcheries" instead.
— Aldous Huxley, in *Brave New World*

We are the children of our landscape; it dictates behavior and even thought in the measure to which we are responsive to it.
— Lawrence Durrell, in *Justine*

Throughout history, employers have selected, or screened, job applicants according to certain hereditary physical characteristics that are valued in the workplace — from body size and visual acuity to gender or a flawless smile. Democratic societies have, of course, enshrined certain legal guarantees in an effort to ensure that this selection process does not degenerate into discrimination against workers because of their skin color, facial features and other inherited traits that are irrelevant to job performance.

In some cases, employers have consciously or unconsciously selected their employees on the basis of visible hereditary traits that they thought would render workers less vulnerable to work-related hazards or illnesses. During the early part of this century, for example, if you sought employment in a tar and creosote factory and happened to have been born with a fair or freckled complexion, your application might have been turned down simply because of your skin color. But the motive for the rejection would not have been racist. Since the late nineteenth century, it has been known that workers constantly exposed to tar, creosote and certain other petroleum products in their jobs were more likely to develop skin cancer than others. And workers with pale skin seemed to be at even greater risk.

In the competitive world of business, one can quickly translate an increased likelihood of contracting occupational disease into dollars and cents. Employee illnesses can mean missed work days, higher turnover of staff, increased training costs and, in recent years at least, spiraling expenses for job benefits and even disastrous lawsuits. So it is hardly surprising that some employers, motivated by what to them were sound financial reasons, simply refused to hire workers who seemed to have visible signs of an inherited vulnerability to skin cancer.

But simplistic attempts to prejudge a worker's ability to perform a job on the basis of his or her genetic constitution, or genotype, pose a potential threat to all of our rights. At what point, one might ask, should a mere statistical probability that a job applicant is genetically vulnerable to a particular occupational disease be considered grounds for denying work to that individual? Should economically motivated judgments about the relative value of human hereditary traits ever be permitted to override deeply held democratic principles of individual freedom and equality of opportunity? And should not every human being be granted a measure of privacy and freedom of choice in important medical matters involving that most personal of sanctuaries — the human genome?

An Early Example of Genetic Screening

Genetic screening can be defined as the examination of the genetic constitution of an individual — whether a fetus, a young child or a mature adult — in search of clues to the likelihood that this person will develop or transmit a hereditary defect or disease. Today sophisticated laboratory techniques in molecular genetics can detect minute differences in DNA sequences. This has made possible the screening of entire populations of job applicants and workers in an effort to identify individuals who might be especially susceptible to particular occupational hazards or illnesses.

The notion of genetically screening workers is not entirely new. Since the early 1960s, tests capable of detecting inherited metabolic differences between individuals have been available. Researchers subsequently began to employ variations of these techniques to search for clues to the genetic basis of occupational diseases. But early attempts were often misguided.

Sickle-cell anemia — a genetic disease that is explored in greater detail in chapter 8 — is a devastating hereditary blood disorder, found almost exclusively in black populations, that can be traced to a defect in a single gene. Homozygotes, who inherit two copies of the mutant gene from their parents, suffer from painful, often life-threatening symptoms of sickle-cell anemia . Heterozygotes, who inherit one mutant and one normal gene, are considered to have sickle-cell trait. Because the majority of their red blood cells function normally, they show no clinical symptoms of sickle-cell anemia. Nor is there reliable evidence that blacks diagnosed as having sickle-cell trait are more likely to suffer under hazardous work conditions than those without the trait.

Nonetheless, the temptation to screen employees for sickle-cell trait, before its possible effects on work performance had been adequately explored, proved too great for the U.S. Air Force

Academy. Fearing that even a single copy of the sickle-cell gene might interfere with the oxygen-carrying capacity of Afro-Americans exposed to high altitudes during pilot training, the Air Force Academy excluded them from flight school over a 10-year period, without adequate scientific evidence to substantiate its concerns. In 1981, under pressure of a lawsuit and unable to demonstrate that low-oxygen conditions would precipitate a medical crisis in these heterozygous carriers, the academy reversed its discriminatory policy.

Yet the absence of compelling scientific evidence that this genotype affects the performance of workers has still not deterred some corporate employers. They have continued to require that black job applicants be screened for sickle-cell trait simply on the assumption, still unproved, that heterozygous carriers of the sickle cell gene *might* turn out to be slightly more vulnerable to oxidizing industrial chemicals, heavy physical exertion, or other job-related stress on their circulatory systems.

More disconcerting, some employers have failed to apply the tests for sickle-cell trait equitably to all populations. Typically, only black applicants are subjected to mandatory screening — even though the sickle-cell gene does occur, albeit rarely, in Caucasian populations. This practice underscores a basic problem with many forms of occupational genetic testing. Since these tests often target genetic variations found disproportionately in certain racial or ethnic groups, they can easily appear to foster job discrimination. Unless occupational tests are developed to detect variant genes in the full racial and ethnic spectrum of worker populations and are equitably applied to the entire work force, job-related genetic screening is likely to inflame workers' perceptions of corporate paternalism and prejudice.

Basic Strategies of Genetic Screening

There are two basic strategies for spotting genetic abnormalities in the human genome. In principle, these can be used to analyze the genetic makeup of cells belonging not just to working adults but also to individuals at virtually any stage of human development. The first strategy depends on the biochemical analysis of substances in the body to reveal the presence of variant genes indirectly.

Tay-Sachs disease, a severely disabling hereditary disorder of the nervous system that occurs in about 1 out of 6000 children of marriages between northern European Jews (versus about 1 out of 600,000 births in marriages of non-Jews of European descent) is one example of such an inborn metabolic error. Two copies of the recessive gene cause a deficiency in the enzyme hexosaminidase, resulting in the damaging buildup of fatty deposits on nerve cells — a process that can be detected biochemically in fetal fluids. Tests are now available that can identify approximately 200 such inherited metabolic illnesses — a small fraction of known genetic disorders — as well as a number of the major chromosomal abnormalities, with the help of detailed family genealogies of hereditary defects. Most such forms of hereditary disease remain both rare and incurable. So while we can screen people for the presence of variant genes, we are often unable to provide effective medical treatments for those diagnosed as having the disease.

The second basic strategy of genetic screening involves the direct examination of chromosomal DNA in human body cells to detect defects that range from major chromosomal anomalies like Down's syndrome to minute differences in the DNA sequences of alleles characteristic of diseases like sickle-cell anemia. This approach often makes use of powerful new recombinant DNA and related laboratory techniques that promise to radically redefine our concepts of genetic "normality" by detecting individual hereditary differences that would normally go unnoticed. Among them are restriction-enzyme analysis to systematically dissect chromosomes with chemical "scalpels," radioactive DNA probes to unveil the exact location of a variant gene, and a variety of cloning strategies for isolating and mass-producing copies of selected genes for closer scrutiny.

Because these genetic tests rely increasingly on the direct analysis of human genes and gene products, rather than on family pedigrees and other more cumbersome methods of monitoring hereditary disorders, they offer a new a window on genetic variation. Already they have begun to reveal a degree of variability previously unsuspected. Instead of having simple pairs of alleles governing a particular trait, in the manner of textbook Mendelian genetics, many genes have an array of possible mutant alleles. Each member of such a gene's family of alternative DNA sequences, or *DNA polymorphisms* (*poly* means "many"; *morph* means "forms"), may differ from another by only a base pair or two. Since each polymorphic allele may alter a healthy individual's condition only in certain environments, and then only slightly, together they can create not just a single "normal" genotype but an entire spectrum of genotypes that can legitimately be considered "normal."

To date, recombinant strategies for screening human genes have found only limited medical application. But that situation is rapidly changing. Geneticists around the world are now engaged in a massive effort to sequence all of the DNA in the human genome. Some researchers have suggested that by the end of this century it may be possible to create from these chemical maps of human genes a toolkit of DNA probes that could be used to construct complete genetic inventories of people. By scanning a patient's allotment of 46 chromosomes for every imaginable hereditary flaw or frailty, one could claim to have developed the ultimate clinical test for "genetic hygiene" — the total gene screen.

Practical Genetic Screening Programs

To grasp the ethical issues at stake in occupational genetic screening — a relatively new and not yet widespread arena of human genetic assessment — it is helpful to review other, more conventional genetic screening programs that are already routinely used in many hospitals. For the most part, these programs are designed to recognize serious hereditary diseases at an early stage in human development — in fetuses and infants rather than in mature adults. But because these programs share the ultimate goal of predicting the health of human beings on the basis of their genetic makeup, they raise many of the same ethical issues as occupational genetic screening and provide a useful medical backdrop.

Prenatal diagnosis represents the most commonly used application of gene-screening technology. It involves the testing of fetal fluids and cells from an unborn child while it is still developing in its mother's womb. After the fluid has been extracted from the fetus, several tests can be performed. Chromosomes can be visually inspected for defects. Restriction enzymes can be used to dissect chromosomes, or cloning techniques can be used to mass-produce copies of selected genes for closer scrutiny. And biochemical assays can reveal the presence or absence of critical gene-coded enzymes.

The most common technique for prenatal genetic screening is known as *amniocentesis* — the examination, usually performed between the fourteenth and sixteenth weeks of pregnancy, of a sample of cell-laden amniotic fluid, the watery membrane-bound cushion surrounding the growing fetus (see Figure 7.1).

Other emerging techniques such as *chorionic villi sampling*, which can analyze cells removed nonsurgically from the hairlike membranes enveloping the embryo during the first 10 weeks of pregnancy, are beginning to offer earlier records of human genotypes. In addition, new recombinant DNA probes and restriction enzyme tests are beginning to prove effective in the early diagnosis of a number of classic hereditary diseases — ranging from devastating metabolic deficiencies such as Lesch-Nyhan disease to blood diseases like sickle-cell anemia. It is important to remember that the vast majority of prenatally diagnosed illnesses are incurable. Until effective treatments become available, such tests can offer little more than scientific guidance to inform the decisions of parents who are willing to consider abortion to prevent the birth of a child who could be gravely ill.

The goals of screening newborn infants for genetic diseases are generally very different. If a genetic defect is identified at a sufficiently early stage of a child's life, it is occasionally possible to alter certain environmental factors to minimize the

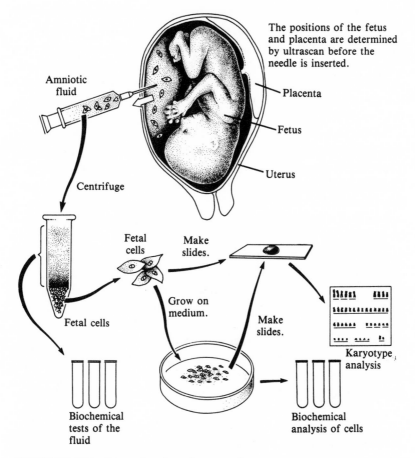

The positions of the fetus and placenta are determined by ultrascan before the needle is inserted.

Amniotic fluid

Placenta

Fetus

Uterus

Centrifuge

Fetal cells

Make slides.

Grow on medium.

Make slides.

Fetal cells

Karyotype analysis

Biochemical tests of the fluid

Biochemical analysis of cells

FIGURE 7.1: Steps in Amniocentesis

damaging impact of that defect. Infants in some countries have been routinely tested for decades, for instance, for an inherited enzyme deficiency known as phenylketonuria (PKU). This condition disrupts the metabolic breakdown, carried out largely in the liver, of the amino acid phenylalanine. Homozygous victims of the recessive disorder — approximately 1 in every 15,000 babies — possess two copies of the mutant PKU gene. As a result, they are genetically incapable of metabolizing this amino acid after eating many protein-containing foods that are common to the diets of most children in developed countries.

The accumulation of unprocessed phenylalanine shifts the metabolic equilibrium in such a way that the amino acid is forced to detour down an alternative biochemical pathway. This diversion produces excess pyruvic acid, a metabolite that can, during the first few months following birth, prove highly toxic to a PKU child's delicate developing brain. Untreated, PKU can eventually lead to brain damage and varying degrees of mental retardation. Thus, early diagnosis is critical to victims of PKU. Once identified, PKU infants can be placed on special low-protein diets that drastically limit their intake of phenylalanine, thereby reducing the buildup of toxic metabolites and allowing brain tissue to develop in a normal fashion.

Genetic testing of adults is generally aimed at either diagnosing a previously unrecognized hereditary ailment or helping prospective parents to prevent the inadvertent transmission of defective genes to their offspring. But it is also the adult population that is likely to be most affected by programs designed to genetically screen employees. In this case, genetic knowledge would not necessarily be used to help individuals decide whether to have children or to proceed with a particular pregnancy. Instead, it tends to have other, quite different purposes.

Genetic Screening in the Workplace

In the workplace, two basic kinds of genetic tests might be used. *Genetic screening tests* are designed to identify individuals who appear to be more susceptible to certain occupational hazards — air pollutants, nuclear radiation, industrial chemicals or other environmental agents — in order to assist managers in making decisions about the hiring and placement of employees. Because these tests inevitably make value judgments about human genotypes that can potentially lead to the denial of a job, transfer or termination, they can be discriminatory and harmful to individual workers.

For example, the red blood cells of people born with a hereditary illness called glucose-6-phosphate dehydrogenase (G-6-PD) deficiency — a disproportionate number of whom happen to be members of racial or ethnic minorities — are thought to be more susceptible to damage from oxygen depletion during prolonged exposure to naphthalene and other industrial chemicals than are the red blood cells of other people. In principle, if the causal connection between the genotype responsible for G-6-PD and job-related anemia were firmly established, voluntary genetic screening programs could ensure that employees at risk would receive expert medical counseling and be given the option of less hazardous work. Used prematurely, the same test might unfairly discriminate against workers because of their variant genotypes.

Genetic monitoring tests, in contrast, are designed not to distinguish the genotypes of individual workers but to detect possible damage to the genetic molecules of workers as a result of chemicals, ionizing radiation or other environmental mutagens that might be found in the workplace. Some genetic monitoring tests analyze workers' chromosomes directly for evidence of occupational abnormalities. For example, a disproportionate number of sperm cells of male workers exposed to an industrial chemical called 1,2-dibromo-3-chloropropane (DBCP), known to cause sterility in men, have an extra Y chromosome resulting from nondisjunction during meiosis. The surplus of sex chromosomes can be rendered visible under the microscope by treating the sperm cells with a special fluorescent dye. Other tests rely on laboratory animals, which are deliberately exposed to potentially hazardous, work-related substances, to serve as sentinels for future human health risks.

By documenting DNA damage in human workers directly or indirectly, genetic monitoring tests can serve as early-warning systems to protect the genetic health of the entire population of workers, rather than to dictate the occupational fate of a few. If genetic tests are eventually devised that can reliably identify workers who are genetically most at risk for developing serious occupational illnesses and if these tests are used in an ethical manner, everyone stands to benefit. Genetic screening programs could prove economically beneficial not only to employers and employees but also to life insurance companies, medical insurance companies, government health plans and other groups or plans that try to anticipate the health risks of working men and women.

The Perils of Occupational Genetic Screening

Despite the immediate short-term benefits of job-related genetic screening, there will always be a number of grave risks. For the most part, they arise not so much from our newfound abilities to instantly analyze human genotypes but from our haste to put these techniques into practice in large human populations before we have grasped their full medical significance.

The technologies underlying new genetic screening tests have improved so rapidly that they have already far outpaced our ability to interpret their results. This imbalance is compounded by the revolution in computer technology. Powerful computer systems permit scientists to rapidly sift through a huge quantity of data on human genetic differences — much of it highly tentative — in search of possible statistical associations with health and disease. As a result, these tentative scientific findings often become available to employers long before the scientific community has had sufficient opportunity to thoughtfully assess them.

Biological "Fate"

Geneticists are devising scores of new tests to detect minute genetic differences in the DNA of otherwise healthy workers that seem to be associated with a predisposition to certain occupational diseases. These statistical associations between specific DNA sequences and the incidence of the disorders are not necessarily causal. It may be, for example, that an individual possessing a particular DNA sequence is not suffering directly from a hereditary illness but rather has inherited some sort of resistance — or susceptibility — to environmentally induced damage.

A lifeless list of every DNA sequence in a person's genome

might seem to some to be an almost prophetic vision of that person's future. But genes represent only one dimension of a person's biological identity — the biologically inherited "nature" component of the eternal "nature-nurture" equation that is acted out in all living things. By focusing exclusively on hereditary factors, even a future "total gene screen" of an individual's DNA can offer little more than a momentary glimpse of the multitude of forces at play in each unfolding human life.

Furthermore, individual differences in DNA often reveal statistical risks for occupational disease that are inconsequential compared with the routine health risks faced by workers every day at home or on the job. A genotype that makes a worker 10 times more likely to suffer an occupational disorder that only occurs in 1 out of every 10,000 people may be intriguing. But this genetic Achilles heel may, in fact, pose far less risk to a person than does commuting to a factory in busy city traffic.

Without worker involvement in the decision-making process, genetic screening tests — many of which have not yet been confirmed as reliable indicators of future health problems—could eventually create arbitrary categories of "genetic acceptability" among workers and job applicants.

Taken to a gloomy extreme, obsessive genetic screening of employees could one day even result in a Huxleyan hierarchical caste system of workers. The lowest rung of the ladder would be occupied by those whose genetic test results marked them as hypersusceptible workers, stigmatizing them as economic untouchables destined to be chronically unemployed. At the highest rung would be workers whose test results established them as model employees whose genotypes — genetically resistant, in one way or another, to the environmental or psychological stresses of important occupational tasks — would guarantee them permanent, if monotonous, positions in the work force.

Problems with Tests

Another question is whether or not a particular genetic constitution will always turn out to be harmful to a worker's health.

Today, geneticists have no doubt that the human gene pool harbors thousands of mutant DNA sequences — most of them yet to be discovered — each of which could conceivably render an individual slightly more susceptible to a particular chemical agent or air pollutant or to another work-related environmental insult than someone with a "normal" genotype. But a test result merely documenting the presence of such a genetic trait does not necessarily mean that the trait will eventually prove pathological to that worker.

Alpha-1-antitrypsin (AAT), for instance, is a gene-coded substance in human blood serum that regulates the activity of special protein-splitting enzymes released by white blood cells — preventing them from accidentally digesting healthy cells. It has been suggested that workers who possess defective AAT genes that result in low levels of AAT in the bloodstream may be genetically predisposed to develop emphysema. This degenerative lung disease can lead to severe breathing difficulties, if one is exposed to chemicals, cigarette smoke or other airborne irritants for prolonged periods of time.

Only about one person in several thousand tested is likely to harbor a double dose of the defective gene. These homozygotes do show an increased vulnerability to emphysema — presumably because enzymes from their white blood cells, unchecked by normal quantities of AAT, attack elastic fibers that give the lung its normal resilience. This loss of resilience is thought to render lung tissue more vulnerable to damage by air pollutants. But workers carrying only one copy of the gene — perhaps 5 to 10 percent of the total population — are still diagnosed as having "intermediate AAT deficiency," even though any link between this heterozygous condition and lung disease remains tenuous.

This is not to say that there is no linkage — only that considerable research remains to be done to isolate the AAT gene's effect from a web of other possible hereditary and environmental factors, many of them still unknown to scientists, that contribute to occupational lung disease. In the meantime, of course, every effort should be made to minimize the exposure of all workers to air pollutants.

In the absence of time-consuming background studies of the underlying causal patterns of each disease, genetic screening programs can give the illusion that they are fostering worker health. Instead, by prematurely labeling carriers of a maligned gene "abnormal" long before the true impact of the heterozygous genotype is fully understood, they run the risk of falsely condemning healthy, genetically variant workers. Being labeled in this way could quickly lead to unwarranted personal concerns, unwanted job relocations or, during difficult economic times, even prolonged unemployment because of presumed genetic unsuitability for certain tasks.

Without legal safeguards to protect the confidentiality of genetic screening data, damage from such misleading medical diagnoses could inadvertently be compounded. Information in corporate or government data banks containing genetic profiles of employees might one day be shared by businesses much as personal credit ratings are — and with a corresponding potential for abuse. For by falsely or prematurely classifying workers as genetically unfit and either openly sharing that information with other businesses or failing to guard against information leaks, companies or government agencies could unjustly stigmatize workers for the duration of their careers.

Genetic tests designed to detect workers who are hypersusceptible to certain environmental conditions can also be expected to be highly impersonal. In prenatal genetic screening, the relationship between physician and patient tends to be intimate — offering at least the possibility of careful interpretation of test findings, assessment of critical nongenetic factors, and auxiliary procedures to confirm an initial diagnosis. When thousands of employees are screened en masse with new, often highly experimental genetic tests, that intimacy is bound to be lost. And with it are sacrificed a number of traditional medical safeguards built in to the doctor-patient relationship.

Among them: a patient's right to be fully informed before submitting to medical tests, the privacy of medical records and the imperative, enshrined in the Hippocratic Oath, to ensure that medical procedures are used to help, or at least not harm, a patient. In the process, the results of genetic screening tests could impose harsh, unilateral economic judgments on the natural variability of the human genome and profoundly diminish the power of individuals to make their own health decisions.

When traditional safeguards are lacking, it is hardly surprising that many observers view the motives of corporate genetic screening programs with undisguised suspicion, especially those that focus on genetic variability within long-oppressed racial and ethnic minorities. When individuals are granted little or no control over highly personal genetic information that could be used to seal their economic fates, such data must be considered vulnerable to abuse.

Especially in societies such as ours, characterized by long histories of prejudice against particular minorities, it is not difficult to imagine how occupational genetic screening could lend an air of scientific legitimacy to attempts, conscious or unconscious, to exclude certain categories of workers from jobs or to restrict their access to more rewarding ones.

Despite their limited medical value, the genetic dossiers arising from mass genetic screening of workers could, just like a factory worker's hereditary skin color, be used to segregate genetically "desirable" job applicants from genetically "undesirable" ones. History demonstrates that the most powerful groups in a society often attempt to justify their status by proclaiming the innate superiority of their race, class or ethnic group. At the same time, victims of slavery, racial oppression and poverty have often heard their plight rationalized by claims of their own supposed biological inferiority. The idea that one's fate is somehow biologically determined by one's genes can easily be exploited to add a veneer of scientific legitimacy to social injustices.

Genetically Resistant Workers versus a Cleaner Workplace
Even if a genetic test is found to reliably predict a worker's susceptibility to certain environmental agents and if that information is treated with respect, another fundamental issue remains. Should such findings be used to exclude affected workers from a polluted

workplace, or should they be used instead to compel the employer to reduce the levels of contaminants to a less toxic level?

It could be argued that many occupational hazards — poisonous fumes, radioactive waste materials and noxious industrial solvents, for example — are seldom encountered in similarly high dosages during the course of most people's lifetime on this planet. Therefore it may be unfair to hastily condemn workers whose genotypes reveal not so much genetic flaws as faint, time-dependent vulnerabilities to a steady onslaught of potent synthetic substances that we seldom, if ever, encounter in the natural world.

Evolution lacks the capacity to anticipate the future. Thus, during the course of human evolution, natural selection could not possibly have equipped our species with mutant genes that could cope with the concentrations of environmental toxins faced by some modern workers. In this light, the industrial workplace is a novel evolutionary setting to which humans have simply not yet had adequate time — which would presumably be hundreds of generations — to adapt. And some slight genetic predisposition to an occupational illness ought to be viewed as a genetic defect only if one assumes that a polluted work environment is a legitimate and desirable part of our species' habitat.

One might also reasonably argue that it is the birthright of every human being not only to dwell in a reasonably unpolluted setting but also to work in one. If this supposition is true, it follows that all but the most genetically vulnerable, hypersusceptible worker detected by a genetic screening program might serve society better in an entirely different role. Workers who are only marginally predisposed to job-related disease need not be viewed as genetically unfit for a job. Rather than penalizing them for their perceived defects, we might look on them as warning beacons representing the final boundary beyond which society should no longer condone further environmental contamination of a workplace.

By insisting that employers reduce the levels of toxic substances to levels that could be tolerated by the overwhelming majority of workers — including those with slightly elevated risks for developing a job-related disease — we would be shifting the focus of blame. Genetic screening tests would direct our attention to an employer's environmental neglect — a variable in the economic equation that could be changed — rather than to an employee's genes — by nature a constant that so far remains beyond our powers to alter in an orderly fashion. At the same time, society would be committing itself not to the creation of a genetically hardier, toxin-resistant workforce but to the creation of a cleaner, more healthful workplace.

The Role of Power in Genetic Screening Programs

As in most clashes between human values and new genetic techniques, power is the underlying thread that connects many of the most crucial moral issues concerning occupational genetic screening: power — political, social and economic — to decide precisely which new areas of genetic knowledge are to be applied. In job-related genetic screening, power is a factor from the instant an idea for a genetic test is first conceived by a laboratory-bound scientist to the moment the results of that test are used to exert control over the lives of a tested worker.

Who, for instance, ought to decide which job-related illnesses should be targeted by new genetic screening techniques? Who should decide precisely when a new genetic test is reliable enough to implement in the form of mass genetic screening programs? Who should determine which human populations ought to be subjected to such genetic scrutiny and which ought, despite an equivalent burden of variant genes, to be ignored? Who should decide which genotypes, if any, ought to be banished from the workplace and on what grounds? And, on a broader scale, who should decide whether new genetic technologies ought to be used to adapt the work force to the work environment or to adapt the work environment to the work force?

The Power Game: A Scenario

For any who might have doubts about the role political power and social status can play whenever judgments about genes are imposed on individuals, consider a brief scenario — a reversal of roles between those people who have traditionally advocated occupational genetic screening programs and those who have usually had to submit to them. Imagine that an administrative committee, composed of some of the least powerful members of a population of workers — racial minorities, non-English-speaking immigrants, poorly educated youth — have suddenly been granted full authority to impose a genetic screening program on their more highly paid employers. The goal: to design and implement a test that will weed out "genetically defective" business executives and midlevel managers. These genetic misfits are then to be efficiently eliminated from corporate payrolls in an effort to improve the company's lagging profits.

First, this powerful committee might decide to provide corporate research funds to spur studies into human genetic variation in areas that might affect the job performance of managers. The search might encompass genes thought to influence the development of a wide range of diseases that are approaching epidemic levels in the ranks of executives, including alcoholism, drug abuse, heart disease, sexual dysfunction and mental illness, to name a few. In time, it is likely that ambitious genetics researchers in both private and public laboratories, flushed by the sudden influx of research grants, would find such DNA sequences or at least identify genetic markers that could be used to signal their presence in genetic screening tests.

Later, other researchers might suggest techniques to carry out low-cost screening programs on the chromosomes of these harried executives, who by now would almost certainly find themselves growing increasingly uneasy over rumors of the committee's benevolent plans to improve their genetic hygiene. As soon as these experimental genetic tests began to promise a degree of predictive value for the target occupational diseases, the committee, emboldened by the new scientific findings, might brashly demand that all managers submit to a series of genetic tests designed to ferret out their "bad" genes. Those managers whose genetic profiles revealed any "undesirable" DNA sequences might then be asked — for their own good health, of course, as well as for the economic health of the corporation — either to transfer to a less stressful position in the company or to seek more "genetically suitable" employment elsewhere.

The pool of genetic information on these unfortunates would then be freely shared with other workers' committees controlling other firms, in the hope that epidemics of alcohol and drug addiction, heart disease and emotional disturbances could finally be controlled. No effort would be made to modify environmental factors that might be contributing to the deterioration of these "executive diseases" — excessive work loads, social stresses, dietary practices and so forth. The diligent genetic screening task force would singlemindedly devote its efforts to identifying what they perceived as disease-prone managers and plucking them unceremoniously from the workplace.

The utter improbability of this imaginary role reversal underscores the imbalance of power that traditionally exists between employers and employees in our society. But this thought experiment also reveals the potential for the abuse of genetic knowledge by any special-interest group, regardless of its ideology, whenever information is used to dictate important decisions to individuals, rather than to enlighten their own personal decision-making processes.

Knowledge as Power
In our society, science advances through the efforts of thousands of individual scientists working in isolated laboratories. For the most part, their discoveries are exploited in ad hoc, piecemeal fashion — with no regulatory system to coordinate the nature and pace of scientific application in ways that would protect cherished human values. Instead, the application of scientific knowledge is determined to a large extent by the turbulent, often value-free forces of the marketplace and an overriding economic imperative to profit quickly from new discoveries. In such a society, those individuals who wield little political power and lack the technical expertise to anticipate the threat posed by some new scientific technologies such as work-related genetic screening are often the most vulnerable to abuse.

Still, there is nothing inherently sinister about geneticists scrutinizing DNA molecules for evidence of genetic predispositions to diseases associated with the workplace — or, for that matter, with the home, the hospital, urban neighborhoods or any other human habitat. All human disease — from cholera and dysentery to cancer and heart disease — are the manifestations of a complex interaction of hereditary and environmental forces. In a just world, an assortment of cheap, reliable genetic tests might be used to alert all human beings at an early age to the statistical risk of contracting many common diseases in certain settings.

With that information, people might be free, assuming their economic conditions permitted such freedom, to alter their behavior in ways that would limit their long-term exposure to those pathogens, water pollutants, dietary practices and other identified environmental factors to which they were diagnosed most vulnerable. Used in this manner, genetic screening techniques might make a major contribution to the global prevention of disease. Of course in a just world, the equal distribution of food, essential medical care, sanitation and other fruits of modern technology would have already eliminated many of those obstacles.

Similarly, the predictive power of genetic tests might also be used altruistically to benefit workers exposed to hazardous job conditions. In American industries alone, for example, nearly 400,000 workers contract disabling occupational illnesses each year — from blood disorders and skin inflammations to respiratory ailments and cancers — and an estimated 100,000 die from such diseases. Tests that encouraged those workers who might be genetically susceptible to particular environmental hazards to avoid those hazards could thus prevent injuries and prolong lives.

But the same tests could also be used to hire or fire workers because of their inherited capacity to tolerate high levels of toxic chemicals, radioactivity or other job hazards. Used in this fashion, new genetic knowledge could result in individuals having less, rather than more, control of their health and lives. And unscrupulous companies could save money by relying exclusively on genetically resistant workers instead of committing funds to cleaning up polluted workplaces.

In genetics, the truth of Francis Bacon's aphorism "Knowledge is power" seems vastly magnified simply because of the nature of the search. Genetics, perhaps more than any other branch of science except brain biology, probes deeply into the identity of individual human beings. By venturing into the privacy of human genomes, geneticists are not simply satisfying, as some would insist, their own insatiable scientific curiosities. Whether they recognize it or not, they are also creating new opportunities for others to harness this scientific knowledge — for good or for ill — in ways that will influence the lives of fellow human beings.

Knowing this, each of us must be willing to do more than simply applaud each startling new breakthrough in molecular genetics that is announced in newspaper headlines. We must also be willing to play a part in monitoring those who might seek to use discoveries in genetics for personal, political or economic leverage in the endlessly shifting balances of power that are the inevitable consequence of scientific knowledge and its application.

CHAPTER 8

GENE THERAPY: THE "MORAL DIFFERENCE" BETWEEN SOMATIC AND GERM CELLS

> **GENETHIC PRINCIPLE**
> While genetic manipulation of human somatic cells may lie in the realm of personal choice, tinkering with human germ cells does not. Germ-cell therapy, without the consent of all members of society, ought to be explicitly forbidden.

Biologists have often emphasized the fact that in general the germ cells form an exceedingly independent tissue; the rest of the body, the "soma," is, in this point of view, merely a temporary structure shielding and conserving the potentially immortal germ plasm.
— Alfred Sherwood Romer, in *The Vertebrate Body*

. . . efforts to engineer specific genetic traits into the germ line of the human species should not be attempted.
— Resolution signed by 56 religious leaders, including Protestant, Roman Catholic and Jewish representatives, and eight scientists and ethicists, *Congressional Record*, June 10, 1983

181

In the early 1980s, Richard Palmiter, a biochemist at the University of Washington, and his colleague Ralph Brinster at the University of Pennsylvania carried out a laboratory experiment that dramatically revealed the promise of genetic engineering in mammals. Using a microscopic needle, they injected cloned rat genes — coding for a growth hormone normally secreted by the pituitary gland in a rat's brain — directly into fertilized mouse eggs in a glass dish. Zygotes that survived this brief trauma were then implanted into the oviducts of hormonally receptive female mice that served as surrogate mothers (see Figure 8.1).

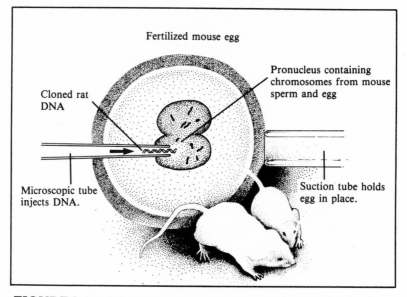

FIGURE 8.1: Microinjection of Rat Gene into Fertilized Mouse Egg

In the end, only a small percentage of the original genetically manipulated eggs developed into mature mice with copies of the recombinant gene for the rat growth hormone. But among these genetically modified, or *transgenic*, animals were mice that had higher levels of growth hormone in their blood and grew to a significantly larger size than their genetically normal littermates.

Thus, these genetically engineered "supermice" had not only incorporated the rat gene into their chromosomes but also expressed the foreign gene by secreting a potent rat hormone into their bloodstreams that visibly accelerated their growth.

In recent years, scientists have successfully injected foreign DNA sequences directly into the reproductive cells or developing embryos of a host of other species — including fruit flies, frogs, rabbits and pigs. These pioneering experiments have generated considerable excitement among geneticists because they open new avenues for exploring the genetics, physiology and development of these species. At the same time, they have also stirred interest in possible practical applications of genetic manipulation in higher animals. Among these applications is the creation of novel, genetically enhanced "superlivestock" harboring foreign DNA sequences designed to spur growth, improve milk or meat production, or perhaps even transform farm animals into genetic factories for the production of hormones, antibodies and other commercially valuable proteins.

What Is Gene Therapy?

In the human sphere, this new technology of tinkering with the genes of reproductive cells and developing embryos has raised hope that we may one day be able to intervene directly to correct genetic errors underlying a number of human hereditary diseases. By demonstrating that foreign DNA can be directly introduced into the cells and chromosomes of animals as complex as mammals, such studies have hastened the day when *gene therapy* — the medical replacement or repair of defective genes in living human cells — will become a reality.

Even the most optimistic of scientists — including researchers already hard at work testing promising gene therapy techniques on chimpanzees and other animal models — are fully aware that gene therapy will not be a medical panacea for our species' genetic ills. Of all the diseases to which humans are heir, only a small fraction has been traced to specific genetic abnormalities —

defects in one or more genes or chromosomes. Of these, only about 3,000, most of them exceedingly rare and often affecting fewer than one in tens of thousands of births, seem to be *monogenic*, or traceable to the activity of a single gene, and are transmitted in simple Mendelian fashion. Among them are such hereditary conditions as albinism — the absence of melanin, the brown pigment responsible for skin, hair and eye coloration; cystic fibrosis — a devastating childhood disease caused by the malfunction of mucus-producing glands in the body; and galactosemia — a metabolic deficiency that blocks the digestion of certain types of sugar molecules.

The vast majority of human diseases — from cancers and viral infections to mental illnesses and dietary deficiencies — seem to be the result of more complex interactions involving more than one gene and a multitude of environmental factors. Yet when talk turns to the future potential for human gene therapy, only monogenic disorders are now likely candidates. And the number of people afflicted by them is minuscule compared with the pressing health concerns faced by most of the world's population.

In spite of these limitations, the notion of extending the ancient battle against disease right into the human genome by directly repairing or replacing defective human DNA sequences represents a radical change in the treatment of hereditary illness. For even the relatively primitive gene therapy techniques we have today portend an enormous new power — unmatched by any previous medical technology — over forces of biological inheritance in humans that have traditionally been ascribed to fate.

Strategies for Gene Therapy
There are several possible strategies for human gene therapy:

1. *Gene Insertion*. The most straightforward approach to gene therapy would be *gene insertion*, which involves the simple insertion of one or more copies of the normal version of a gene into the chromosomes of a diseased cell. Once expressed, these supplementary genes could produce sufficient quantities of a missing enzyme or structural protein, for example, to overcome the inherited deficit.

2. *Gene Modification*. A more delicate feat, *gene modification,* would entail the chemical modification of the defective DNA sequence right where it lies in the living cell, in an effort to recode its genetic message to match that of the normal allele. In principle, this approach would be less likely to disrupt the intricate geographical layout of genes on chromosomes, thereby reducing the possibility of unwanted side effects.
3. *Gene Surgery*. Even more daring would be bona fide *gene surgery* — the precise removal of a faulty gene from a chromosome, followed by its replacement with a a cloned substitute. The notion of transplanting freshly synthesized DNA sequences to replace defective ones is the ultimate dream of gene therapy. But it would demand a mastery of human genetics that, for the moment at least, simply exceeds the geneticist's grasp.

Eventually, all three approaches may prove medically feasible. But only the first — gene insertion — appears to be technically feasible for the immediate future. At this writing, we have no record of the successful genetic treatment of a person with an inherited disease. But, with a number of prominent research groups increasingly eager to begin clinical trials of gene-transfer techniques on human volunteers suffering from rare, hopelessly incurable genetic diseases, that may soon change.

The diseases that appear to be the most promising targets for these pioneering experiments are so rare that their names are unfamiliar to most of us. Most are recessive genetic diseases — those in which the homozygous patient receives one defective gene from each parent. Unlike dominant diseases — those in which a single copy of the abnormal gene, inherited from only one parent, is sufficient to trigger symptoms — these ought, in theory, to respond positively to the insertion of a single normal copy of the gene.

One such recessive disease is Lesch-Nyhan syndrome, a devastating neurological disorder that can cause cerebral palsy, mental retardation and even a tragic, uncontrollable urge to mutilate one's own body. It affects about one person in 10,000, most of them children. Like many so-called inborn errors of metabolism caused by single-gene defects, it can be traced to a critical missing enzyme. Its absence effectively blocks an essential biochemical reaction within nerve and other body cells, resulting in a buildup of toxic levels of unprocessed intermediate chemicals — in this case, uric acid.

Two other candidates for gene therapy, adenosine deaminase (ADA) deficiency and purine nucleoside phosphorylase (PNP) deficiency, can be traced to an absence of enzymes that in turn cripples the body's immune system. No more than 100 cases of both these diseases have been reported throughout the world. But their victims, rendered genetically defenseless from birth, are ravaged by wave after wave of life-threatening infections. In the more distant future, other, more familiar hereditary illnesses may also prove genetically repairable. These include hemophilia — the blood-clotting disorder that afflicted Queen Victoria's royal pedigree; Huntington's chorea — which eventually crippled folksinger Woody Guthrie; and, probably long after other monogenic disorders have been mastered, even certain polygenic diseases such as diabetes.

A Scenario for Gene Therapy in Sickle-Cell Anemia

No single imaginary scenario for gene therapy can reveal all of the possible risks and rewards that might arise from the future medical manipulation of human genes. Nor could such a scenario claim to be scientifically accurate in all of its details until successful experiments have been carried out on human subjects. But we offer a hypothetical case involving sickle-cell anemia, also known as sickle-cell disease, that illustrates the basic biological principles that are likely to be at work in future attempts at human gene therapy — and the implications of applying them on a large scale.

Sickle-cell anemia is a rare but geographically widespread genetic blood disorder, found almost exclusively in black populations in Africa or among their descendants elsewhere in the world. The symptoms of sickle-cell anemia include insufficient levels of hemoglobin in the blood, or anemia; impaired circulation, leading to local damage to internal organs; episodes of excruciating pain in bones and joints; and a reduced life span.

The disease affects the lives of millions of people around the world, and 1 in 500 black infants in the United States is born with the homozygous blood disorder. With such frequencies, it is hardly surprising that many African tribes in Ghana and elsewhere have long been keenly aware of the disease and its familial pattern of inheritance — if not its precise cause.

The Discovery of Sickle-Cell Anemia

Even so, Western medicine did not discover the disease until the twentieth century. In 1910, while examining a blood smear from a black West Indian medical student, Chicago physician James B. Herrick first observed the clusters of irregularly shaped red blood cells that are the hallmark of sickle-cell anemia. Viewed through a microscope, a normal human red blood cell has the shape of a smooth round pill, centrally indented, or concave, on both surfaces. But many of this patient's red blood cells were strikingly angular and elongated — resembling the curved blade of a sickle or scythe (see Figure 8.2).

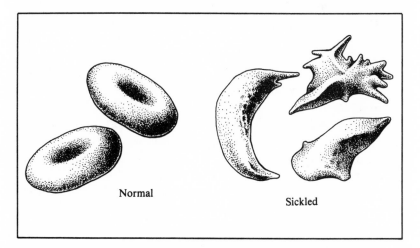

Normal

Sickled

FIGURE 8.2: Normal and Sickled Red Blood Cells

Because Herrick happened to be more interested in diseases of the heart than of blood, he viewed the abnormal cells with little more than passing interest. But others soon noted that the novel sickle-cell blood disorder seemed to be transmitted along family lines and that the crumpling, or sickling, of the red blood cells was reversible. Simply by increasing oxygen levels in the blood one could force the sickle-shaped cells to swell back up into biconcave disks resembling healthy red blood cells. By reducing oxygen concentrations again, one could force them to collapse back into sickle shapes once more. Furthermore, the curious cells could be found in the blood not only of patients suffering from debilitating anemia but also — though in limited quantities — in patients who seemed perfectly healthy.

The Genetic Basis of Sickle-Cell Anemia

It was not until the late 1940s that details of the underlying genetic basis of sickle-cell anemia began to unfold. By studying the family pedigrees of patients, geneticists discovered that the disease arises from an error in a single gene and is transmitted from parent to offspring as a recessive characteristic. In homozygous double dose — one gene from each parent — the disease floods the bloodstream of a victim with flawed red blood cells that buckle into sickle shapes as soon as they encounter sufficiently low oxygen concentrations in the blood. As they career through the circulatory system, these deformed cells have a tendency to impair blood circulation by rendering it more viscous. They also become lodged in narrow capillaries, creating local obstructions that could eventually result in swelling, bouts of severe pain, and damage to oxygen-starved tissues. In the heterozygous condition known as sickle-cell trait, the combination of one abnormal gene and one normal one still leads to the production of some defective red blood cells — but not in quantities sufficient to trigger the devastating clinical symptoms of sickle-cell anemia.

In 1949, the Nobel Prize–winning American chemist Linus Pauling and his colleagues demonstrated that the disease could be traced to a structural defect in a single molecule resident in red blood cells — hemoglobin. This defect creates an abnormal protein known as *S* (for sickle-cell) hemoglobin. The discovery not only opened up a new chapter in the story of sickle-cell anemia but also established an entirely new category of human illness — molecular disease. By immersing hemoglobin proteins in a gelatinous substance and exposing it to a weak electrical current — a process called *electrophoresis* — Pauling and his colleagues demonstrated for the first time that normal hemoglobin and *S* hemoglobin were chemically distinct. Because the two hemoglobin variants differ slightly in electrical charge, they are propelled through the electrical field at a different rate, thereby creating two separate pools of protein.

By determining the exact order in which amino acids in the two proteins were linked together, other researchers identified their exact structural differences. Virtually all of the hemoglobin molecules in the red blood cells of a normal adult, it turns out, are built of four interlinked chains of amino acids called *globins*. Two of them — each 140 amino acids long — are identical polypeptide chains called alpha globins. The other two — each 146 amino acids long — are identical beta globins. A pair of duplicate genes, located on human chromosome number 16, code for the two alpha polypeptides. Another gene, located on chromosome number 11, codes for the two beta peptides.

After the four polypeptide chains have been synthesized, they join together to form a single folded, four-lobed protein. Around the same time, a quartet of small, highly reactive iron-bearing molecules called *heme groups* bond at strategic locations within the convoluted, three-dimensional structure. Each heme group is specially equipped to form transient chemical bonds with oxygen atoms. As red blood cells carrying heme molecules pass through the lungs, the heme molecules snatch up oxygen atoms, only to release them later when the red blood cells circulate through oxygen-hungry tissues elsewhere in the body.

Magnified and placed side by side, the structure of a molecule of normal hemoglobin and that of its genetically defective kin — *S* hemoglobin — are almost identical. If you compared them

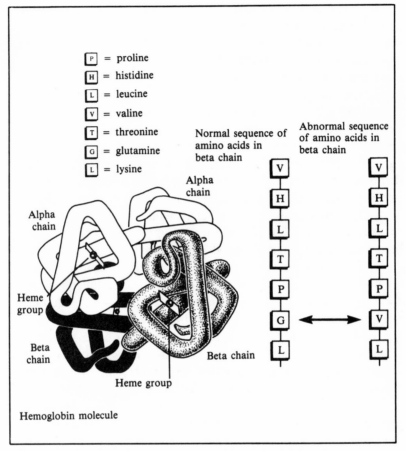

FIGURE 8.3: Normal and S *Hemoglobin Molecules*

— amino acid by amino acid — you would be unable to spot the difference until you reached the sixth amino acid of a beta chain (see Figure 8.3). There you would discover the fatal flaw responsible for sickle-cell anemia: a single substitution of one amino acid for another. In position number six, *S* hemoglobin has an amino acid named valine in place of the required amino acid called glutamic acid. This tiny blemish — one out of hundreds of precisely linked amino acids — can be traced to an error in a single nucleotide in the gene coding for the defective beta chain. But that is enough to transform the nucleotide triplet, or codon, GTC (guanine-thymine-cytosine), specifying glutamic acid in normal hemoglobin, into the codon GTG (guanine-thymine-guanine), specifying valine in S hemoglobin. And it is enough to render the entire amino acid framework of the huge protein vulnerable to untimely collapse each time blood oxygen levels decrease.

This might seem a trifling molecular mistake to trigger such life-threatening symptoms. In many cases, in fact, the substitution of a single amino acid in the formula of an enzyme or a structural protein molecule may have little or no impact on its ability to function in the cell. But in hemoglobin, the link between its architecture and oxygen-transport activity is so intimate that even a change in one amino acid can have catastrophic consequences.

When the amino acid valine is switched with glutamic acid, the three-dimensional shape of the hemoglobin molecule is subtly altered. The instant a red blood cell containing mutant S hemoglobin encounters low concentrations of blood oxygen, the distorted hemoglobin molecule begins to behave eccentrically. It combines with adjacent hemoglobins, creating a rigid crystalline structure that causes the cell to collapse. As the sickled cell reenters more oxygen-rich regions of the circulatory system, molecules disengage, allowing the red blood cell to swell into its familiar disklike shape.

There is no effective treatment or long-term cure for sickle-cell anemia. Especially in lesser-developed nations, it is possible to improve the outlook of patients somewhat simply by providing adequate nutrition, sanitary conditions and basic medical care. But, despite decades of research, there remains nothing in the vast medical pharmacopoeia that can safely and reliably control the sickling of red blood cells. For now, physicians can do little more than try to diminish the effects of sickling by using antibiotics to control secondary bacterial infections or painkilling drugs to cope with the inevitable, agonizing bouts of pain.

Recent Medical Advances

In recent years, the most dramatic advances in the medical management of sickle-cell anemia are related to early detection and prevention. Using recombinant DNA techniques, geneticists can now diagnose the disease in the first months of embryonic development from amniotic fluid, without the trauma of extracting a fetal blood sample. Early diagnosis allows parents who have already witnessed the painful suffering of their previous offspring the option of choosing therapeutic abortion if they wish.

In addition, adults who have been diagnosed as having either sickle-cell anemia or sickle-cell trait can receive genetic counseling concerning the risks of transmitting the disease to their children. If both partners are diagnosed as having the heterozygous sickle-cell trait, they can be informed that on average 25 percent of their children will be born with homozygous sickle-cell anemia. If one partner has sickle-cell anemia and the other partner has only sickle-cell trait — one normal gene for the beta chain in hemoglobin and one abnormal one — probability dictates that 50 percent of their children will be homozygous for the defective gene and suffer from sickle-cell anemia.

Because sickle-cell anemia is painfully debilitating, easily diagnosed, incurable and burdensome not only to those individuals who suffer from it directly but also to future generations who will inevitably inherit it, it is an obvious candidate for gene therapy. It seems perfectly reasonable to assume that a remedy might lie simply in providing red blood cells with a copy of the correct beta-chain gene. But the fact that the hemoglobin molecule is the product of more than one gene could complicate the situation. During the assembly of two pairs of identical chains in each hemoglobin molecule, the activities of alpha and beta influence one another. The sudden intrusion of additional copies of one gene, creating a shift in the relative quantities of alpha and beta chains, could disrupt the fragile interplay between genes. So it is conceivable that attempts at gene therapy could even prove more harmful than healing.

For this reason, several less complicated, single-gene diseases

— among them, the aforementioned Lesch-Nyhan syndrome, ADA deficiency and PNP deficiency — have emerged as more promising targets for the first pioneering efforts to repair the human genome genetically. But the basic strategies for their genetic treatment are unlikely to differ drastically from those we present for sickle-cell anemia.

One Recipe for Gene Therapy

The first applications of gene therapy techniques on human subjects are likely to focus exclusively on body, or somatic, cells. For sickle-cell anemia, this means trying to manipulate genes of somatic cells located in the blood-forming tissues of human bone marrow.

The reason why gene therapists would attempt to manipulate DNA sequences in the bone marrow cells of a patient with sickle-cell anemia rather than in circulating red blood cells is simple. Red blood cells routinely lose their gene-laden nuclei as they mature. The average life span of a circulating red blood cell is approximately 120 days — with each blink of your eyes, a million or more blood cells die. The body's supply of red blood cells is continuously replenished through mitotic cell divisions in the blood-forming precursor or stem cells lodged in the cavities of your skull, ribs, vertebrae and other bones. These actively dividing cells each possess a full complement of genes and thus, in people with sickle-cell anemia, harbor the genetic error responsible for the disease. In principle at least, one ought to be able to improve the prognosis of these patients — perhaps even cure them completely — simply by restoring the correct DNA sequences in these genetically defective blood-forming cells.

Step 1: Isolation of the Normal Gene

Gene therapy for sickle-cell anemia is likely to require four basic steps (see Figure 8.4). In the first step, the gene coding for a normal beta chain in hemoglobin would be identified and physically isolated from the cells of a healthy human being. This step has already been carried out using standard gene-cloning techniques — resulting in laboratory cultures of bacterial cells harboring millions of identical copies of the critical donor gene.

Step 2: Removal of Defective Cells from Patient

In the second step, bone marrow cells harboring the defective beta-chain gene would be removed from the patient to facilitate the genetic exchange. It is easier and far less risky to genetically manipulate human cells in laboratory cultures of cells under controlled conditions than to navigate genes to a specific cluster of cells in a living patient who harbors tens of trillions of vulnerable cells. In fact, the former approach has already been carried out on a variety of laboratory animals.

Fortunately, the bone marrow cells that are the source of the patient's genetically defective red blood cells are quite accessible. Living bone marrow cells can be removed from bone cavities using a syringe or another surgical instrument with minimal damage to surrounding organs. In fact, such procedures are now a routine part of bone marrow transplant operations in which cancerous blood-forming cells in a patient suffering from leukemia, for example, are replaced with cells from a healthy, tissue-matched donor.

Because of the relative ease of such operations, most of the hereditary illnesses now being considered as candidates for gene therapy are blood and immune disorders affecting bone marrow tissues. The treatment of hereditary diseases involving brain, heart and other more fragile tissues is likely to prove more daunting. Presumably it will rely on innovative techniques for transporting replacement genes directly to target cells in the body, instead of simply removing cells for laboratory manipulation of their defective genetic molecules.

Step 3: Insertion of the Normal Gene into Cells

The third step in gene therapy for sickle-cell anemia would be to insert copies of the desired DNA sequence into living bone marrow cells in such a way that they perform their desired function.

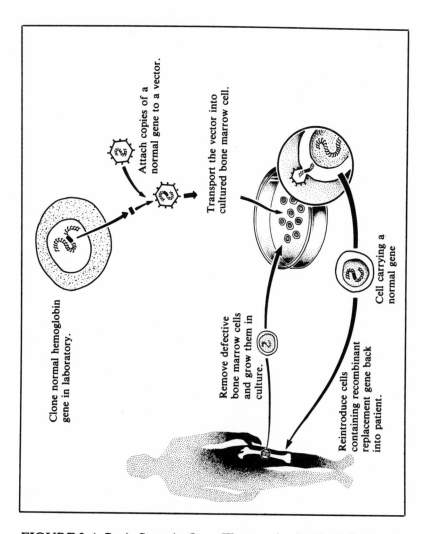

Clone normal hemoglobin gene in laboratory.

Attach copies of a normal gene to a vector.

Transport the vector into cultured bone marrow cell.

Cell carrying a normal gene

Remove defective bone marrow cells and grow them in culture.

Reintroduce cells containing recombinant replacement gene back into patient.

FIGURE 8.4: Basic Steps in Gene Therapy for Sickle-Cell Anemia

The actual transport of the cloned beta-chain genes might be carried out in a variety of different ways. For example, genes can be laboriously injected into a living cell by hand, using a fine glass tube, or micropipette, to puncture the cell's outer membrane and squirt a solution of DNA into its fluid interior. It is also possible to soak target cells in certain chemicals that encourage the cells to absorb foreign DNA directly. Alternatively, small synthetic sacs known as *liposomes* can be filled with cloned genes and then fused with the membranes of target cells, allowing these DNA sequences to drift into the cells. Or genes can be spliced to the genetic molecules of certain species of viruses, which, during the course of their natural infection of cells, can catapult cloned DNA sequences directly into the chromosomes of the target cell.

Step 4: Reintroduction of Cells into Patient's Body
The fourth and final step in the genetic treatment of sickle-cell anemia would be to transplant the genetically modified bone marrow cells back into the patient's body. The cells — each equipped now with at least one newly inserted copy of the hemoglobin gene coding for a normal beta chain — could be injected directly into the victim's bone cavities, just as in a conventional bone marrow transplant operation.

As the transplanted cells flourished, the expression of foreign genes would lead to healing flows of molecularly correct beta chains that would be incorporated into the hemoglobin assembly lines of each descendant cell. Then, when mature red blood cells, laden with normal hemoglobin molecules, were finally released into the bloodstream to carry oxygen, they could function without sickling.

Possible Medical Risks of Human Gene Therapy
This scenario for treating sickle-cell anemia seems reasonable enough on the surface. But a number of subtle biological obstacles must be overcome before gene therapy on human patients could ever become a routine medical procedure.

Gene Expression
It is not enough simply to insert replacement genes into a defective genome. Before the gene therapy operation can be considered successful, the transplanted DNA sequence must be expressed — that is, it must be transcribed into messenger RNA, then

translated into biologically active polypeptide products in the bone marrow host cell. Not only must gene expression be achieved, the gene must perform its task without unduly disrupting other vital metabolic processes and it must remain physically integrated in a chromosome during subsequent cell divisions.

The prospects of gene expression are greatly enhanced by splicing specific promoter regions — DNA control sequences that regulate the activity of other known genes — to the imported beta-chain gene. This technique permits the replacement gene to be properly transcribed into messenger RNA and, in turn, into a chemically correct chain of amino acids. Certain promoter sequences permit the expression of a gene only in specific body tissues. By assigning such tissue-specific promoters to a transplanted gene, one can guarantee that a particular gene will be expressed exclusively in those organs of the body — the brain, liver or pancreas, for example — in which the missing gene product is required.

Precision
A number of other obstacles must be overcome before the genetic manipulation of somatic cells could become a reliable medical treatment. For one thing, geneticists cannot yet begin to control the precise chromosomal destination of genes sent into a target cell. For that matter, they cannot even be sure of the quantity of cloned copies that will be delivered.

With current delivery techniques, the genes seem to integrate into the cell's chromosomes randomly — with all the surgical finesse of pellets from a shotgun blast. And the risk of randomly inserting genes may be far greater than one might guess. For example, the haphazard addition of genes, as previously mentioned, could conceivably distort the intricate chemical interplay between members of a gene family dedicated to the synthesis of a desired metabolic product.

In genetically modified bone marrow cells, for instance, a newly inserted beta-chain gene would have to interact harmoniously with a pair of alpha-chain genes to synthesize functional hemoglobin molecules and to maintain stable levels of hemoglobin in the bloodstream. The addition of foreign DNA sequences into these cells could disrupt the body's finely tuned, time-tested, genetic regulation of these processes.

Early indications are that gene therapy is more likely to lead to an inadequate supply of a missing gene product rather than a surplus. But it is possible that by tinkering with globin genes one might inadvertently program them to synthesize either too little or too much hemoglobin. And, given the uncertainties of the trajectory of a transplanted gene, the medical consequences of one such genetic operation might differ dramatically from the next. In contrast, individuals suffering from more severe, life-threatening hereditary diseases might be willing to assume such medical risks for even the slight possibility of increasing the quantity of a missing enzyme or structural protein.

Genetic Damage
The haphazard trajectories of transplanted genes could also have far more hazardous consequences. Randomly inserted into a genome, a gene — alone or in multiple copies — can land in the middle of an existing gene, shattering the continuity of its linear DNA sequence. In experiments with laboratory animals, for example, such events — called *insertional mutations* — arise fairly frequently.

To date, the majority of these genetic changes seem to result in recessive rather than dominant mutations. So they do not necessarily represent a health hazard to the genetically modified somatic cells. But if reproductive cells in the organism inadvertently suffered similar mutations, the consequences could conceivably prove more sobering. In an egg or sperm cell, even a recessive insertional mutation might pass unnoticed for a time, only to surface unexpectedly as a homozygous trait in the body cells of a developing embryo or its descendants.

In addition, scientists have recently learned that some human genes have the capacity, under certain conditions, to transform the normal growth pattern of a cell into a cancerous one. Among the events that can incite a perfectly normal human gene to turn into a renegade, cancer-causing *oncogene* are a sudden change in the identity of neighboring genes and an infection with certain viruses — including members of the same retrovirus family of viruses that hold such promise as gene-shuttling vectors in the medical transfer of genes. Until gene therapy strategies can promise absolute accuracy in the targeting of genes, they run the risk of unintentionally igniting cancers in treated cells — some of which might not be be visible in patients for years.

Thus, in much the same way that the miracle drug insulin has simultaneously saved the lives of millions of diabetics and also created unexpected long-term health problems for many patients, it is possible that gene therapy could provide tantalizing short-term health benefits in exchange for unknown, long-term health costs. Because the first patients to volunteer for somatic-cell gene therapy trials are likely to be desperately ill individuals suffering from fatal hereditary diseases, the risk of future genetic abnormalities or cancers may be of little concern. But in the more distant future, enthusiastic gene surgeons may well be tempted to use gene therapy techniques on a broader scale to treat less catastrophic disorders. In this event, the potential hidden health consequences of meddling with the human genome could not be ignored.

The "Moral Difference" between Somatic and Germ Cells

It has been argued that the potential risks of human gene therapy are in many ways no different from those of any other innovation in the tradition of Western medicine. Desperately ill patients volunteering for experimental genetic treatments, so the argument goes, have a right to assume risks in the hope of improved health — just as others willingly submit to experimental cancer drugs, daring new brain operations or unconventional organ transplants. And, as long as gene therapists confine their genetic manipulation to somatic cells, bounded by the finite life span of an individual patient, this argument is difficult to refute.

In this light, the application of gene therapy techniques to the blood-forming somatic cells of a person suffering from sickle-cell anemia differs little from the use of bone marrow transplants for victims of leukemia. In both cases, cells with genotypes that differ from the patient's genotype are imported to correct a life-threatening genetic disorder. Thus, gene therapy can be seen as the ethical equivalent of an organ-transplant operation — a local transformation in a patient's phenotype, without a corresponding change in the underlying genotype of his or her reproductive cells.

But the ethical dimensions of gene therapy shift the instant the target changes from somatic cells, with fleeting lifetimes, to germ, or reproductive, cells belonging to lineages that are potentially immortal. The fact that early proposals for future human gene therapy experiments have focused exclusively on somatic cells suggests that most scientists recognize the profound biological difference between somatic and germ-line cells in the human body. But, in part, this early reluctance to tinker with the genes of human germ cells genes can also be attributed to the imprecision of current gene therapy strategies. As the medical art of gene therapy becomes more refined, we must be on guard against those who may be inclined to apply these skills to the germ-line genetic engineering of our species. For, coming on the heels of successful somatic-cell repair, proposals for germ-line gene therapy are likely to be cloaked in the noblest of medical motives — the dream of saving lives in future generations and ultimately even eradicating hereditary disease.

The Paradox of "Good" versus "Bad" Genes

Consider once again the case of sickle-cell anemia. If gene therapy one day proves effective in the treatment of bone marrow cells, eventually the temptation to seek a permanent "cure" for the disease by repairing germ-line cells may prove overwhelming. In principle, if normal beta-chain genes were inserted into egg or sperm cells of patients with sickle-cell anemia, these individuals would not transmit the defective sickle-cell gene to their progeny but would instead produce genetically normal offspring. On a more practical level, germ-line therapy might also dramatically reduce long-term health care costs for the disease. For by repairing a patient's defective germ-cell genome once, it seems, a gene therapist would avoid having to repair the somatic-cell genomes of that patient's descendants in every successive generation.

Unfortunately, there is a problem with this tantalizing plan; the sickle-cell gene is not as "bad" as it might at first seem. Early studies of the distribution of the sickle-cell gene in tropical Africa showed that in some areas as many as 30 percent of the native populations were heterozygotes with asymptomatic sickle-cell trait. One would not expect the genotype for sickle-cell trait genotype to be so prevalent unless it were somehow beneficial to human survival. What possible evolutionary advantage could the possession of a heterozygous genotype with one normal beta-chain gene and one defective one confer on its owner?

Equally perplexing, the geographic distribution of populations with high frequencies of sickle-cell trait tended to coincide with regions having a high incidence of malaria. The agent of one variety of malaria is the microbial parasite *Plasmodium falciparum*, which is injected into the human bloodstream by the bite of a bloodthirsty female mosquito. At one stage in its life cycle, this malarial parasite invades circulating red blood cells, where it reproduces. Swollen with parasites, the red blood cells burst, triggering periodic bouts of fever and chills that are the hallmark of the disease.

But, paradoxically, people with sickle-cell trait have a better chance of surviving the battle with malaria than those who harbor either two copies of the mutant sickle-cell gene or two normal genes. The key to the success of these heterozygotes seems to lie in their genetically determined quota of defective red blood cells. When these cells are attacked by the parasites, they sickle, just as they would in low-oxygen conditions. But when the parasite attacks, the collapse proves beneficial. As the body's immune system is mobilized to track down and remove the accumulating debris of sickled red blood cells, it simultaneously sweeps up millions of the malarial parasites trapped inside them. As a result, the parasitic infection is often thwarted and the lives of many individuals with sickle-cell trait are spared.

The message in the selective advantage of the heterozygous sickle-cell trait condition is not that the sickle-cell gene ought to be forever preserved in the human gene pool, like some rare species of endangered animal. It would be difficult to make a case that black populations in malaria-free North America, for example, should be left to endure the agonies of sickle-cell anemia simply to preserve a natural genetic defense against potential future epidemics of malaria.

But the story of sickle-cell anemia does underscore the striking capacity of a seemingly defective gene to simultaneously offer both advantages and disadvantages — depending on its quantities and surroundings. Moreover, there is growing evidence that natural selection often routinely maintains a balance of such mutant DNA sequences in populations of many species. The problem is that we are blind to that fragile balance except in rare instances, such as sickle-cell anemia, where geneticists have managed to unravel the intricacies of gene action in relation to hereditary disease.

But how many other "defective" genes responsible for hereditary disorders might harbor some unseen evolutionary value? And, until we know enough about human genetics to begin to grasp their evolutionary roles, what price might we eventually pay if we overzealously try to "cure" these genetic abnormalities by ridding the human gene pool of these DNA sequences using germ-line gene therapy?

The Feasibility of Germ-Line Gene Therapy

The prospect of human germ-line genetic manipulation is hardly science fiction. For evidence of this, we need look no farther than the pioneering gene-transfer experiments in rodents described at the beginning of this chapter. In those experiments, Palmiter and Brinster microinjected growth hormone genes from rats into the fertilized eggs of mice. A portion of the genetically modified mouse embryos that survived transplantation into surrogate mothers developed into visibly larger supermice — genetically engineered mice that revealed the presence of the rat DNA sequences by synthesizing growth-stimulating flows of rat hormone in their bodies.

But an equally important result — and one that is easily overlooked — was the discovery that some individuals in this first generation of supermice transmitted their extraordinary new phenotype to their descendants. The reason that the trait for large size was inherited by the offspring of these transgenic mice is not difficult to understand. Because the rat genes were introduced directly into newly fertilized eggs and integrated into resident chromosomes before the cells had undergone the first mitotic cell divisions of embryonic development, copies of the rat DNA sequence were subsequently distributed to every cell in the mouse's developing body — including its embryonic reproductive cells. Thus, many sexually mature mice passed the foreign gene on via their genetically altered egg or sperm cells.

The significance of this finding is profound. A rat gene had been forced to traverse the biological boundary between somatic cells and germ cells in another species. It had soared from mortal soma of a single mouse to the immortal germ plasm shared by all genetically related laboratory mice. And this event had occurred in a mammal — a species whose mode of sexual reproduction and embryonic development closely resembles our own.

The micromanipulation of fertilized eggs or early embryos is not the only means by which new genes could be inserted into human germ cells — and, thereby, into the human gene pool. Virtually any system for delivering genes that has proved successful for somatic-cell gene therapy — virus vectors, liposomes, chemical shock treatments — could, with certain modifications, hold promise for germ-cell therapy. Crucial to the application of any of these approaches to human germ-line therapy is a reliable system for scrutinizing and manipulating the decisive genetic events of human sexual reproduction — the fusion of egg and sperm and the early mitotic cell divisions of the developing embryo — in the laboratory.

Newspaper headlines trumpeting medical breakthroughs with test-tube babies, surrogate mothers, sperm banks and frozen embryos remind us that this art of external, or in vitro, fertilization is already a reality. Remarkably, even as society becomes increasingly concerned about the potential social and ethical effects of these and other human reproductive technologies, we hear little public discussion about the role these same techniques may play in future human germ-line therapy procedures. The fact is that almost any scientific advance that grants greater access to human reproductive cells and embryos hastens the day when the genetic manipulation of human germ cells will become medically feasible. Each new breakthrough in human reproductive technology, in a sense, draws the human egg, sperm and embryo one step farther out of the dark, protective sanctuary of the womb and into the bright light of future genetic control.

The Perils of Germ-Line Gene Therapy

Underlying the altruistic aims of germ-line therapy lies the potentially dangerous perception that a troublesome gene can somehow deliberately be erased from the human gene pool. The history of eugenics suggests that once a human characteristic — such as a particular skin color or mutant hemoglobin molecules or poor performance on IQ tests — has been labeled a genetic "defect," we can expect voices in society to eventually call for the systematic elimination of those traits in the name of genetic hygiene. But the notion that the human genome is like some sort of genetic garden from which hereditary defects can simply be plucked like so many weeds is both mistaken and dangerously naive.

First, any genetic disease can only be provisionally defined. A gene that fails to perform to our satisfaction under one set of nutritional, climatic or other environmental conditions might possibly perform quite satisfactorily in another setting. In this sense, a gene's "defectiveness" can sometimes be a transient quality. And, simply by identifying the nutrients or environmental conditions necessary to sidestep the harmful effects of a mutant gene, one might be able treat some hereditary disorder without recourse to genetic intervention.

Second, there is a continuous upwelling of new disease-causing DNA sequences in the human genome as a result of random mutations. Genetic imperfection is an unavoidable characteristic of human hereditary processes; it is part of what makes us human. And, as the source of both the raw genetic variability on which natural selection depends and the inevitable errors that sometimes result in genetic disease, that imperfection is always double-edged.

Third, to try to eliminate a recessive mutant gene causing a rare genetic disorder, one would have to treat not only the occasional homozygous individuals who suffer from the disease but also asymptomatic, heterozygous carriers harboring a single heritable copy of the same allele. This would require massive genetic screening programs to sift through the genotypes of the entire human population, the vast majority of whom were healthy. A less ambitious program aimed strictly at repairing the genes of homozygous individuals and thereby culling their defective alleles from the gene pool presents other problems. The decline in the frequency of the mutant gene would proceed at a ponderously slow rate. As the frequency declines, however, the cost of diagnosing each case would show a corresponding rise. So inefficient would such a negative eugenics program be that it might take hundreds of generations and many centuries to achieve a meaningful decrease in the frequency of a deleterious recessive gene.

Fourth, even if the germ cells of someone suffering from a hereditary disease could somehow be genetically patched, other defective genes, recessive alleles temporarily masked by dominant ones, would continue to be transmitted. It has been estimated that every human being carries between five and 10 hidden defective genes, so that even individuals who appear perfectly healthy can contribute to the incidence of genetic disease.

Finally, and perhaps most important, the overwhelming majority of inherited human traits — intelligence, height and skin pigmentation, for instance — tend to be polygenic. That is, they arise from the interplay of more than one gene and are characterized by a continuous rather than a steplike Mendelian variation in phenotype. For now, at least, such multigene targets remain beyond the reach of those who might consider employing new gene therapy techniques as a tool to shape their particular dreams of human genetic hygiene. Because a polygenic trait deemed "defective" arises, by definition, from the interplay of a combination of genes, purging it from the gene pool would often require impossibly complex repairs. Even assuming these repairs could be carried out, one could never be sure that altering a constellation of polygenes would not inadvertently disturb other, seemingly unrelated cellular processes influenced by these genes.

The Future of Gene Therapy

The idea of seeking a long-term cure for a genetic disease by introducing new genes directly into human germ-line cells is nothing less than revolutionary. It diverges from a long tradition in mainstream Western medicine of focusing therapy on somatic cells. Germ-line therapy, in sharp contrast, would grant physicians the power to prescribe treatment not simply for a single ailing patient but for all of his or her progeny.

Medicine has never completely ignored germ cells. By providing genetic counseling to victims of inherited disorders — for example, offering parents the option of medically aborting their genetically defective fetus or even encouraging the compulsory sterilization of mental patients — physicians have often shown a willingness to influence the fate of human germ-line cells. But until recently they have lacked the technical skills to tinker with individual genes in germ cells or to assign genes to unborn generations, which are incapable of expressing informed consent.

There is no reason to fear the stunning new conceptions of human hereditary disease now emerging from genetics research. In fact, we can rejoice that this new genetic knowledge is certain to improve the prevention, detection and treatment of many previously untreatable genetic disorders. At the same time, each of us shares responsibility for ensuring that techniques allowing the manipulation of the human genome are never exploited for arbitrary and self-serving ends or in ways that fail to consider the potential long-term consequences of large-scale genetic repair on human populations.

In the application of gene therapy techniques exclusively to human somatic cells — cells of the bone marrow, liver or brain, for example — we may be protected from our mistakes by the mortality of these cells. Since the life spans of somatic cells do not exceed those of the human organism, any sacrifice arising from errors or miscalculations in genetic repair is likely to be limited to the loss of life of the patient undergoing genetic treatment. Even with this built-in biological buffer zone, we should not grant permission to our healers to commence surgery on the human genome until society has had ample time to debate its possible effects on a whole spectrum of human values — including our most fundamental definitions of what it means to be human.

But in the future, if gene therapy techniques are ever aimed at human germ cells, the cost of our ignorance of genetic processes would instantly become unacceptably high. Genetic manipulation of human reproduction has the potential to multiply medical errors exponentially — sending ripples that radiate far beyond the finite lifetimes of gene therapist or consenting patient. Our ethical judgments, individually and collectively, ought always to reflect this profound biological difference between somatic cells — with their short-lived genes that lie in the moral domain of individual choice — and germ cells — with their potentially immortal genes to which future generations also lay moral claim.

CHAPTER 9

BIOLOGICAL WEAPONS: A DARK SIDE OF THE NEW GENETICS

> **GENETHIC PRINCIPLE**
> The development of biological weapons is a misapplication of genetics that is morally unacceptable — as is the air of secrecy that often surrounds it.

> *. . . the emergence of the new biology requires that the entire subject of biological weapons disarmament be reopened. . . . One window upon Apocalypse is more than enough. . . .*
> — Susan Wright and Robert L. Sinsheimer, in *Bulletin of Atomic Scientists,* November 1983

> *It is deemed unethical for doctors to weaken the physical and mental strength of a human being without therapeutic justification, and to employ scientific knowledge to imperil health or destroy life.*
> — *Code of Ethics in Wartime,* World Medical Association

Biological warfare can be defined as the deliberate use of microorganisms or toxic substances derived from living cells for hostile purposes — that is, to kill, injure or incapacitate human beings or the animals or plants on which they depend. It has aptly been called "public health in reverse," for it is founded on this dark premise: that the very pathogens — disease-causing viruses, bacteria, fungi and other microorganisms — against which medicine has waged endless battle can be used to military advantage by harming the health of political foes.

In principle, almost any agent causing a disease known to harm human beings, their crops or their livestock could be used to fashion some sort of biological weapon. But in practice relatively few of these organisms satisfy the basic military prerequisites for a biological warfare agent. In general, these pathogens must be suited to mass cultivation in factories, be able to withstand artificial modes of storage and dispersal (e.g., in bombs or aerosol sprays) and cause rapid, predictable outbreaks of disease in target populations without harming attacking troops. Even with these practical restrictions, nature still offers biological warfare enthusiasts a bounty of potential biological weapons.

Any list of the militarily most promising diseases usually includes a number of humanity's most feared scourges, along with an assortment of more minor, often mildly debilitating diseases (see Table 9.1). The names of some of the most virulent of these read like a litany of plagues from the past. Among the most terrifying of them are anthrax, a hardy, bacterial infection affecting both humans and domestic animals; pneumonic plague, a respiratory form of the ancient flea-borne bubonic plagues of the Middle Ages; smallpox, a fatal viral illness; and yellow fever, a mosquito-borne viral disease of the tropics.

The idea that science might *deliberately* try to transform humankind's shared reservoir of medical knowledge into a potent military arsenal is morally abhorrent to most people. Yet while they might be quick to acknowledge the possibility of subtle, indirect hazards inherent in many applications of genetics, few feel the need to contemplate the prospect of darker, manifestly malevolent schemes to harness genes. However, few seem to realize that even as we celebrate pioneering medical breakthroughs arising from recent developments in recombinant DNA

TABLE 9.1
Selected Biological Warfare Diseases

DISEASE	PATHOGEN	SYMPTOMS
Diseases Caused by Bacteria or Bacterialike Agents		
Anthrax	*Bacillus anthracis*	Inhaling spores leads to respiratory form — fever, respiratory distress, shock in 3-5 days; fatal.
Brucellosis	*Brucella melitensis*	Long-lasting fever, weakness, aching; rarely fatal.
Cholera	*Vibrio cholerae*	Sudden, severe intestinal disease — diarrhea, vomiting; sometimes fatal if untreated.
Glanders	*Actinobacillus mallei*	Infectious disease of horses, mules, donkeys and occasionally humans.
Plague	*Pasteurella pestis*	Severe fever, shock, delirium; respiratory form very infectious and often fatal.
Q Fever	*Coxiella burnetii*	Debilitating fever, chills, weakness; rarely fatal.
Tularemia	*Pasteurella tularensis*	Severe fever; pulmonary form estimated to cause about 40% fatality. (Note: Tularemia was the standard American lethal biological agent until Richard Nixon announced in 1969 that U.S. biological weapons stockpiles would be destroyed.)

Diseases Caused by Viruses

Dengue fever	Dengue viruses	Sudden fever, headache, joint and muscle pain; rarely fatal.
Smallpox	Pox virus	Sudden fever, malaise, skin lesions; often fatal in unvaccinated individuals.
Venezuelan equine encephalitis	VEE virus	Sudden headache, fever, malaise; occasionally fatal. (Note: VEE was the standard American incapacitating agent until 1969.)
Yellow Fever	Yellow Fever virus	Sudden fever, nausea, jaundice; potentially fatal.

Diseases Caused by Fungi

Coccidioidomycosis	*Coccidioides immitis*	Inhaling spores leads to fever; rarely fatal.
Wheat blast and rice blast		Highly virulent, viral plant diseases. (Note: Wheat and rice blast were the standard U.S. anticrop agents until 1969.)

Toxins	Source
Botulinus	*Clostridium botulinum* bacteria
Ricin	Toxic protein from castor bean seeds
Saxitoxin	Poison produced by marine dinoflagellates (genus *Gonyaulax*)

technology and molecular genetics, there are indications that at least some scientists in the United States, the Soviet Union and other nations have received military funding to search for ways to use this revolutionary knowledge to design new, more sophisticated components — all of them ostensibly "defensive" — to existing biological weapons systems.

We would be naive to ignore the possibility that some of the same genetic techniques that are now finding so many promising and beneficial applications in modern medicine — new vaccines, cloned hormones and other pharmaceutical products and prenatal diagnostic tests for disease — could be applied to blatantly harmful, "antimedical" ends. In the the wrong hands, these techniques might, for example, be used to create arsenals of genetically modified microbes that could ignite uncontrollable epidemics in unwary civilian populations or to concoct proprietary vaccines reserved exclusively for the use of military troops or friendly civilian populations. Military applications of genetic technologies are every bit as feasible as medical ones. All that is required is money, scientific expertise, an atmosphere of moral indifference and the same sort of ingenuity that fuels practical advances in other, more humane applications of modern genetics.

Early Use of Biological Weapons
Biological warfare is an ancient military strategy, although one that has rarely been used. Despite the veil of secrecy that has traditionally surrounded the subject, historians have traced it back thousands of years to civilizations as diverse as the Greek, Roman and Song Chinese. One of the earliest recorded instances of biological warfare occurred in 600 B.C. when the Athenian legislator Solon poisoned the water supply in the city of Kirrha with the noxious roots of the *Helleborus* plant — a primitive, but nonetheless effective, biological toxin. During the fourteenth century, a Tartar army, while laying siege to the Crimean town of Kaffa, now part of the Soviet Union, catapulted the bodies of plague victims over the walls of the city to ignite an epidemic in the enemy population. And, during the eighteenth century,

British soldiers, and later U.S. government Indian agents, deliberately infected North American Indian populations by providing them with trade blankets previously contaminated with the deadly smallpox virus.

But only in the past century have biological weapons begun to receive serious consideration as a legitimate mode of large-scale warfare. In part, this development can be traced to a series of scientific discoveries in cell biology, bacteriology and microbiology that took place in the nineteenth and early twentieth centuries. By allowing biologists to identify, isolate and culture pathogenic microbes in the laboratory, these discoveries opened the door to the medical diagnosis, treatment and prevention of many of the infectious diseases that had plagued humanity throughout history. But that same knowledge also enabled people to more efficiently manipulate certain infectious diseases for military ends.

World War I is better known for introducing mass chemical warfare to the battlefront than for any innovations in biological warfare. During that war, opposing armies drenched each other with more than 100,000 tonnes of poison gas — first chlorine, then phosgene and finally mustard gases — blistering, suffocating or otherwise disabling an estimated 1.3 million soldiers and causing nearly 100,000 deaths. But there were also disturbing allegations, difficult to substantiate, that military strategists were beginning to incorporate biological weapons into their arsenals. In 1915, for example, the United States accused German agents of secretly inoculating U.S. shipments of horses and cattle from Argentina with infectious bacteria. And, over the next two years, similar attempts were said to have been made to infect European livestock with anthrax and glanders, which can also infect human beings.

In response to growing public concern over chemical and biological warfare in the wake of World War I, a clause was incorporated into an international treaty, the 1925 Geneva Protocol, specifically forbidding "bacteriological methods of warfare." But other ominous developments in chemical and biological warfare would continue to take place. In 1936, for instance, while investigating commercial insecticides for private industry, German researchers concocted a recipe that was a precursor to a new generation of more deadly chemical weapons. Bearing benign commercial trademarks like Tabun, Sarin and Soman, this family of paralytic nerve gases disrupts communications between nerve cells in its human victims — in much the same way that similar products affect insect pests. It has been suggested that Hitler came close to unleashing Germany's vast stores of nerve gas during the final desperate campaigns of World War II, but refrained, in part, out of fear of possible Allied retaliation.

At the same time, Allied forces were hoarding their own stocks of toxic gases. Official documents indicate that Churchill seriously contemplated deploying these against German cities in aerial attacks. In the end, he refrained, official documents suggest, largely because of tactical concerns.

World War II

Not until World War II had ended would it become known how earnestly both Allied and Axis — particularly the Japanese — forces had pursued the development of biological weapons. British, Canadian and American scientists, for instance, pooled their scientific expertise at government biological weapons research centers — Porton Down in England; Suffield, Alberta, not far from Medicine Hat; Camp Detrick, Maryland; and elsewhere — to carry out research and conduct open-air field tests of some of the most promising pathogens. Anthrax, with its hardy, long-lived and highly infective spores, turned out to be the Allied disease of choice. Before the war had ended, the United States and Britain had plans in place to manufacture thousands of 2-kilogram anthrax bomblets, designed to fit into cluster bombs that weighed over 200 kilograms and exploded in midair. The bombs were supposed to have been assembled in a factory in Indiana, but the war ended before production hazards could be resolved. The British did succeed in stockpiling approximately five million anthrax-laced cattle cakes designed to infect enemy livestock populations.

But it was the Japanese who carried out the most serious program in biological weapons research and development during World War II. It was centered at a research station in the small Manchurian village of Pingfan, near the town of Harbin. Established in 1937, the Harbin facility represented the world's first major biological weapons center and at its peak housed 3,000 scientists, technicians and soldiers — all dedicated to devising biological bombs. The diseases they investigated included typhoid, typhus, botulism, plague, cholera, brucellosis and anthrax. They reportedly cultivated up to seven tonnes of bacteria per month in their laboratory vats, detonated thousands of experimental anthrax bombs and destroyed thousands of laboratory and domestic animals to document the effects of candidate diseases.

The Harbin scientists went so far as to deliberately infect prisoners of war with a number of lethal diseases. In some cases, these human subjects were tethered to stakes and exposed to flying fragments from specially designed bombs previously contaminated with anthrax spores. Even a minor shrapnel wound from such a bomb usually proved fatal. The exact toll of these grisly experiments is unknown. But several hundred captive prisoners — mostly Chinese, but also British, American and Australian soldiers — are thought to have been killed in this "scientific" quest to refine the ancient arts of biological warfare. Even more disturbing are suggestions that the United States and other nations may have enhanced their own biological weapons programs by importing Japanese expertise shortly after the war.

Post-World War II

There have been numerous allegations of the actual release — accidental or intentional — of biological weapons during the decades since World War II. But for the most part these allegations remain poorly documented, hotly debated and subject to the inevitable distortions of propaganda campaigns waged by superpowers eager to justify their own "defensive" biological weapons research programs by publicly portraying their adversaries' programs as "offensive." Over the years, for example, the United States has been accused of deliberately trying to infect North Korean populations with plague bacilli during the early 1950s and of igniting costly viral epidemics among the domestic livestock of Castro's Cuba in the 1960s.

The Soviet Union, for its part, has been blamed for incidents such as the sudden outbreak of anthrax within its own borders in 1979 — allegedly arising from an accident in a biological munitions factory in the town of Sverdlovsk — and the appearance of fungal toxins — known popularly as Yellow Rain — in war-torn Laos, Cambodia and Afghanistan. Reputable scientists, however, have since pointed out that the initial reports of fungal toxins could not be confirmed using more rigorous methods of chemical analysis. And the Yellow Rain itself has been found to be composed mainly of pollen and to be nothing more than a harmless, yellow material excreted by flights of local honey bees.

It is not our intention to attempt to untangle the complex web of scientific, legal and political issues involved in these and other accusations. But their very existence reminds us that the potential usefulness of biological weapons — both on the battlefield and off — has not escaped modern military minds.

Further evidence to support this conclusion can be found in rising military budgets for biological defense systems, which soared from approximately U.S. $15 million in 1980 to about U.S. $70 million in 1987. According to the National Science Foundation, the U.S. Department of Defense allotted U.S. $100 million for biological research in 1983. By late 1983, it was also openly funding research projects in its biological weapons program in the private biotechnology industry and in academic institutions. These projects explored such topics as recombinant DNA vaccines for diseases of military interest. Among them: Rift Valley fever and dengue fever, two mosquito-borne viral illnesses common to tropical regions that incapacitate victims with sudden bouts of fever, headache, malaise and other symptoms; Q fever, a tick-borne disease, caused by rikettsia, bacterialike microorganisms slightly larger than viruses and triggering similar symptoms; and anthrax — easily the most deadly of the four.

None of these diseases poses a significant threat to the health of the American public, although some of them can be considered a threat to the health of U.S. troops stationed abroad. All happen to be prime candidates for biological weapons. In 1985, the U.S. Army reported spending $39 million on defensive biological weapons programs. The Soviet Union is reported to be funding biological weapons programs of its own, although, of course, detailed information is unavailable.

Equally unsettling, the federal Recombinant DNA Advisory Committee (RAC) in the United States, the final guardian of scientific responsibility concerning federal recombinant DNA research funded by the National Institutes of Health, has approved gene-splicing projects that are clearly of interest to the military. They include the cloning of genes coding for such biological poisons as exotoxin A, secreted by a small rod-shaped soil bacterium called *Pseudomonas*; a neurotoxin released by *Shigella* bacterium, a major cause of dysentery in Third World countries; and the deadly botulinum toxin responsible for some forms of food poisoning. The stated rationale for permitting the cloning of such toxins has been to create poison-laden antibodies that can be directed against cancerous cells as a possible medical treatment — a worthy goal. But critics have argued that safer ways may exist to achieve these and other similar research goals. In any case, many perceive a need for RAC and other recombinant DNA regulatory institutions around the world to explicitly forbid any projects that are militarily motivated.

The Military Pros and Cons of Biological Bombs
What advantage could superpower military agencies, their arsenals already bristling with nuclear and conventional weapons, possibly seek in developing new, more efficient biological weapons? There may be several advantages — though they have not yet proved compelling enough to transform this traditionally unconventional form of warfare into a conventional one. In the first place, biological weapons are potentially more selective — though not necessarily accurate or precise — in causing destruction than

flames, explosives or bullets. By zeroing in on living organisms, they can weaken or destroy enemy forces without damaging valuable physical property. Because of nature's incredible reservoir of diseases, biological weapons also offer a wide range of military responses, from swift, surgical guerrilla assaults on human populations using incapacitating agents to fungal epidemics that sweep through thousands of hectares of food crops. Compared with nuclear weapons, for example, biological weapons are also extraordinarily cheap, easy to mass-produce — they are self-replicating — and portable. In this sense, they can be seen as a potential low-budget deterrence system for poorer nations that cannot afford to build their own nuclear weapons arsenals.

But their most seductive feature may be their suitability for clandestine attack. Even genetically engineered microbes may turn out to be virtually indistinguishable from natural forms. Because microorganisms require time to reproduce and initiate infection, biological weapons also possess built-in time delays. The biological equivalent of fuses, they can provide precious hours, days, even weeks for a terrorist or saboteur to escape — making it even more difficult to trace a sudden epidemic to its source.

There are also distinct disadvantages. Foremost among them is the simple fact that most people around the world are morally repulsed by the idea that the military organizations of any nation might exploit deadly human disease. For this reason, the political consequences of negative world opinion in the wake of a documented biological warfare attack could easily outweigh any short-term military advantages.

Another disadvantage is the inherent unpredictability of any military arsenal that relies on biological systems. The genetic variability in populations of organisms, the inexorable process of gene mutation and the incalculable effects of environmental variables — from wind, weather and air pollution — all combine to guarantee an element of tactical uncertainty in launching an attack with a biological weapon. To this can be added the ever-present fear that the effects of a disease might boomerang, inadvertently killing or incapacitating the very troops that launched the attack — not to mention unsuspecting civilian populations living nearby. Not even amassing stockpiles of vaccines or antibiotic drugs against potential biological warfare agents is likely to fully lay this gnawing fear to rest. The wide range of natural pathogens available as biological weapons include many agents for which we currently have no effective vaccine or pharmaceutical remedy. Even more disturbing, recombinant DNA technologies now raise the specter of new strains of novel, genetically modified microbes controlled by military scientists. Such creations could eventually make the notion of an impenetrable defensive shield against future biological attack little more than a dangerous illusion.

Like any novel, genetically engineered organism concocted in the laboratory, new generations of recombinant biological weapons could conceivably cause long-term environmental consequences that are beyond our scientific ability to calculate. North American ecologists have already documented numerous historical cases in which the mere introduction of an exotic species — from the Asian fungus responsible for igniting epidemics of chestnut blight in eastern forests to flocks of aggressive European starlings that displaced the local species of songbirds — has led to significant changes in North American ecosystems.

But even with their dazzling, computerized models of ecosystems, ecologists are simply unable to predict the ecological impact of releasing waves of military microbes, each with its own novel genotype, around the globe. Would a novel pathogen, by causing low-level infections in nontarget species, for example, create new reservoirs of disease that would become a permanent feature of the biological landscape? Would it interact with other naturally occurring pathogens to produce a new disease that no one could possibly have anticipated? Would the pathogen evolve in ways that might quickly render existing methods of military controls obsolete — thereby triggering uncontrollable epidemics that could linger on to infect future human generations long after the original motives for the biological attack had been forgotten?

"Offensive" versus "Defensive" Weapons

However complex, a nation's motives for engaging in biological weapons research today may, in fact, have little to do with any balance sheet of military pros and cons. Biological weapons programs can be fueled by forces ranging from the political posturing of leaders for critics at home and adversaries abroad to the narrow self-interest of ambitious scientists working in the biological weapons field. Understandably, the single most compelling factor can be a nation's fear that a foreign rival might secretly be hard at work using new scientific strategies to design its own state-of-the-art, genetically engineered agent, against which the former nation possesses not even rudimentary countermeasures. It is precisely to reduce such uncertainties that existing international treaties restricting the development of biological weapons must be respected, nurtured and strengthened.

Invariably, governments insist on referring to such programs as purely "defensive." Their use of recombinant DNA technologies in biological weapons research, they claim, is restricted to searching for new military shields — genetically engineered vaccines, portable detection devices, diagnostic tests — in preparation for possible enemy attack. But there is a growing awareness today that the distinction between "offensive" and "defensive" biological weapons is so blurred as to be almost meaningless.

In the first place, vaccines and other "defensive" biological armaments are as essential to a complete, functional biological weapons system as are "offensive" ones. In the same way that a spear-throwing warrior — deprived of his traditional shield — would hesitate in launching an offensive attack, a modern military establishment lacking expertise in defensive measures against biological warfare agents could not realistically initiate an offensive attack.

In the second place, experts in molecular genetics have noted that many of the technical processes required to develop "defensive" biological arms are inextricably linked to the development of "offensive" ones. To design a new drug antidote for a deadly biological toxin, for instance, one might first require cloned copies of the DNA sequence coding for the poisonous substance so that it could be synthesized in quantity. Or, to develop a genetically engineered vaccine, one might first have to possess the offensive biological agent against which the vaccine was to be prepared. In addition, one would also need to carry out practical tests — using living organisms, perhaps even human subjects, infected with the agent — to determine the effectiveness of the new vaccine. Thus the question arises: Can biological weapons research ever be "purely defensive" when it is so intricately and unavoidably intertwined with "offensive" biological weapons research?

The answer is that it cannot. This does not mean, of course, that a nation's biological research program cannot be motivated by genuine defensive concerns. But one would expect that the scope of such a "defensive" research effort would differ fundamentally from that of a major "offensive" biological weapons program aimed at designing, delivering and deploying a battle-ready arsenal of biological weapons. And the only hope for reliably distinguishing between these two very different government biological warfare policies rests on the total elimination of official secrecy surrounding biological weapons research and full public disclosure of research findings.

The Future of Genetically Engineered Weapons

Because of the atmosphere of official secrecy that surrounds biological weaponry in most countries, it is difficult to do more than speculate on the possible directions that the military application of modern, molecular genetics could be taking. It is fair to assume that virtually *any* approach that grants greater human control over the dynamics of infectious or toxic diseases is likely to be considered.

Genetically Modified Microbes

One possible strategy, for example, might be to genetically modify existing pathogens to create new generations of "supergerms"

— microbes that are even more efficient than natural diseases at killing or disabling selected military targets. Many biomedical experts remain openly skeptical that recombinant DNA techniques can be employed to "improve" on the pathological habits of naturally occurring agents of disease that have been fine-tuned by natural selection. And, for now, they may be right. After all, the bond between pathogen and host represents a delicate balance between the life cycles of two species that could easily be disrupted by any first, crude attempts to manipulate a pathogen's genes.

But, in certain cases at least, the macabre scientific challenge of trying to design a still deadlier or more accurate form of disease may not turn out to be such a daunting feat. We already know, for instance, that resistance to antibiotic drugs in many species of bacteria is mediated by gene-bearing plasmids — tiny ringlets of DNA that not only are routinely exchanged in nature but also can be artificially shuffled in the laboratory. Thus, military scientists might try to genetically engineer bacterial agents that survived dousings of antibiotic drugs used by enemy forces to contain the outbreak of disease in the wake of a biological attack. Similarly, the military might try to genetically modify bacteria so that they could be treated by secret pharmaceutical formulas held exclusively by friendly forces.

In principle, genetic alterations could yield a wide range of other useful "improvements" in a pathogen's ability to cause disease. They might, for example, lead to hardier, longer-lived pathogens — ones better equipped to withstand the rigors of factory production, deterioration during storage and prolonged exposure to sunlight and air as they drift downwards from the sky in aerosol clouds or parachute bombs following aerial attacks. Other "improvements" might include more aggressive microbes — ones armed, for example, with transplanted foreign DNA sequences designed to make them more virulent — or more cosmopolitan microbes — ones capable of invading subtly different areas of a host organism's anatomy or adapting to diverse geographic landscapes and latitudes that would not normally sustain them.

Even more devious, it is possible that someone could try to

create a highly specialized biological warfare agent designed to carry out a single, sinister mission — to disrupt communications within the tightly coordinated armies of cells in the human immune system that devour alien microbes and secrete neutralizing antibodies. The advantage of designing a pathogen that would strike in this fashion is obvious. By first dismantling the human body's built-in, biological barriers to infection, such an agent might instantly expose its host to hordes of local pathogens that constantly surround it but are normally unable to harm an individual with a healthy, intact immune system.

Such strategies would not necessarily lead to a quantum leap in the dangers already posed by existing biological agents or by conventional methods of altering them to increase their military effectiveness. But, in principle, genetic engineering techniques could provide greater precision and new opportunities for innovation in the genetic modification of military pathogens.

Ethnic Weapons

Another diabolical scheme might be to enhance the ability of certain pathogens to home in on highly specific target groups of host organisms. A particularly nightmarish scenario — one that was actually proposed by a scientist writing in the American military journal *Military Review* in 1970 — would be to transform microbes into so-called ethnic weapons, agents of biological warfare that might exploit subtle hereditary differences between human populations.

The idea would be to modify a pathogen so that it would be more likely to infect members of a particular racial or ethnic group while leaving others relatively unscathed. In this way, the military might be able to deploy an offensive biological weapon that, in a sense, embodied its own defense. That is, hereditary membership in the "right" group would provide one with a measure of built-in protection against illness because of the agent's engineered host preference.

Scientists have already begun to accumulate a long list of

human diseases that disproportionately affect members of specific racial or ethnic groups. But there is no documented evidence that effective ethnic weapons have been designed to exploit these. Nonetheless, future insights into the genetic basis of biological differences between human beings are likely to uncover hereditary traits that could be used to military advantage.

Some scientists have speculated, for example, that it might be possible to manufacture pure antibodies that could home in on protein molecules found predominantly or even exclusively in the cells of individuals belonging to particular ethnic groups. These antibodies could then be armed with potent toxins designed to kill any recognized cells on contact.

Alternatively, it might be possible to design a biological weapon that exploits slight variations in human metabolic processes. It is known, for example, that in adult American Indians, Australian aborigines and a number of other racial groups, the enzyme lactase, essential to the digestion of lactose sugar in milk, is rare or absent. Yet most light-skinned people of northern and central European descent continue to manufacture the enzyme throughout their lives. If similar genetic variations could be identified that more perfectly correlate with race, they could conceivably be viewed by military researchers as metabolic Achilles heels that would be vulnerable to attack.

Cloning Biological Toxins

The most practical strategy for biological weapons may prove to be the use of existing gene-cloning techniques to mass-produce biological poisons inside bacteria or yeast host cells cultivated in huge vats in much the same way biotechnologists already clone human proteins such as insulin or interferon. This procedure would allow military scientists to produce enormous quantities of deadly poisons — such as snake venoms, shellfish toxins or bacterial toxins — that are otherwise extremely difficult to harvest from organisms in nature.

One potential candidate for cloning would be the toxin excreted by the microbe causing botulinus food poisoning — the bacterium *Clostridium botulinum.* It is one of the most toxic substances known to humanity; a mere millionth of a gram of botulinus toxin can be enough to kill a human being. The World Health Organization has estimated that less than a kilogram of this substance, placed by terrorists or saboteurs in the water supply reservoir of a typical North American town of 50,000 people, could conceivably kill 60 percent of the population within a period of 24 hours. Others have suggested that the toxin might be far less potent because the fragile protein is easily denatured under such conditions.

Genetically Engineered Vaccines

Another biological warfare strategy — one already being openly pursued by the American military — is the creation of genetically engineered vaccines to immunize troops against possible attack. The steps needed to prepare vaccines against at least some potential biological agents are likely to differ little from those required to construct cloned vaccines we already possess against certain diseases (see Figure 9.1).

The first task is to isolate a gene coding for a microbial substance that triggers an immune response — typically a small protein subunit studding the protective outer coat of a virus or bacterium. This molecule, the critical ingredient in any vaccine, is then mass-produced by cloning the gene inside *E. coli* or another convenient host cell. The direct inoculation of friendly troops with the vaccine could confer immunity against future biological warfare attack with the same pathogen. It has even been suggested that it might one day be feasible to mass-immunize civilian populations, without their knowledge, by releasing an aerosol mist containing a vaccine over a city. This prospect seems highly unlikely in the near future. But the motive behind such a clandestine maneuver would presumably be a desire to protect civilians from an impending attack, without inciting hysteria, despair or moral outrage in the process.

Early-Detection Devices

Genetic and immune techniques used for medical diagnosis might also be harnessed to provide extremely sensitive early-warning

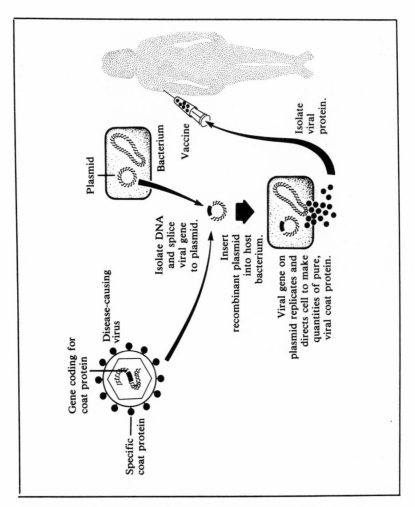

FIGURE 9.1: Steps in Cloning a Human Vaccine

systems for the detection of biological warfare agents. These techniques might be harnessed, for instance, to identify pathogens with customized nucleic acid probes — radioactively labeled DNA sequences designed to home in on and bind to complementary DNA sequences known to reside in the genetic material of a familiar microbe. Similarly, pure, laboratory-synthesized anti-

bodies might be released to zero in on characteristic proteins studding the outer surface of an invading microbe.

In combination with powerful computer software and miniatur-
ized electronic circuitry, these and other sophisticated detection
systems might eventually be reduced to compact monitoring
devices that could continuously scan air flows circulating through
military command centers or wafting over battlefields.

Agricultural Weapons
Practical offensive biological warfare agents might also be
genetically designed to more effectively attack an enemy's
agricultural crops or livestock. Especially in poor, developing
nations of the Third World, where agricultural diseases are often
inadequately monitored because of prohibitive costs, it might well
be impossible to distinguish such an intentional outbreak of
disease in a country's cash crops from an unusually severe natural
epidemic. It is even conceivable that industrialized nations that
provide the bulk of the world's commercial seed for major food
crops could conspire to secretly develop biological warfare agents
tailor-made to attack only those varieties of plants destined for
politically hostile lands. Ironically, though, the growing dependence of these developed nations on genetically uniform seed for
their own monoculture crops of wheat, corn and other staples
could render them especially vulnerable to biological
counterattack.

The Biohazards of Genetically Engineered Weapons
The use of genetically engineered biological weapons has been
prophesied — though seldom publicly discussed — since the
earliest days of the recombinant DNA era. Yet in February 1975,
when over 100 leading scientists in molecular genetics met at the
landmark Asilomar Conference to discuss the potential risks of
gene-splicing experiments, the topic of biological weapons
research was not placed on the main agenda.

The Asilomar Conference is justifiably remembered as a bold
exercise in social responsibility. It exemplified the extraordinary

willingness of scientists to publicly air even their most farfetched concerns regarding the possible consequences of carrying out gene-splicing research in the hope of creating guidelines to reduce the risks to society. Nonetheless, discussion of every major biohazard on the Asilomar agenda shared the premise that the only harmful impact that one could expect from a future recombinant DNA project would result from scientific miscalculation or rare laboratory accidents — never from malicious intent. The possibility that a fellow scientist might one day *intentionally* employ his or her gene-splicing techniques to harm human or nonhuman species was not publicly raised as a serious moral issue.

As a result, the probability that the hazardous scenarios hypothesized at Asilomar would actually happen was usually dismissed as statistically small. But military applications of recombinant DNA technologies could conceivably cast doubt on this conventional wisdom. Hazardous events that might be seen to have no more than a "vanishingly small" (a vaguely defined phrase that tends to be overused by some scientific experts) possibility of taking place inside the laboratory of a responsible research scientist might be transformed into far more frequent occurrences inside a biological weapons laboratory dedicated to enhancing the military usefulness of pathogens (see Table 9.2).

It is interesting to note that prior to the Asilomar meeting, a small committee of experts, members of the Plasmid Working Group, was asked by conference organizers to review the potential risks of artificially exchanging bacterial plasmids that confer resistance to antibiotic drugs. Among them were such highly regarded scientists as Richard Novick, Royston C. Clowes, Stanley N. Cohen, Roy Curtiss III and Stanley Falkow. In their brief written report, the committee members went beyond biohazards strictly related to plasmid exchange to address what they saw as the ominous threat of biological weapons research based on recombinant DNA techniques. While their succinct statement was not openly debated at the Asilomar Conference, its message rings hauntingly true more than a decade later:

We believe that perhaps the greatest potential for biohazards involving alteration of microorganisms relates to possible military applications. We believe strongly that construction of genetically altered microorganisms for any military purpose should be expressly prohibited by international treaty, and we urge that such prohibition be agreed upon as expeditiously as possible.

TABLE 9.2
Potentially Hazardous Recombinant DNA Experiments Discussed at Asilomar Conference, 1975

- Inserting genes coding for toxins into bacteria
- Changing the host range of bacteria
- Using animal viruses as vectors
- Inserting genes coding for drug-related substances into bacteria
- Cloning DNA from animal viruses
- Randomly inserting DNA from higher organisms into lower organisms
- Dispersing antibiotic-resistant plasmids in bacterial populations

(Adapted from Sheldon Krimsky, *Genetic Alchemy: The Social History of the Recombinant DNA Controversy,* Cambridge, Mass.: The MIT Press, 1982, p. 373.)

International Treaties on Biological Warfare
Some experts on biological weapons disarmament maintain that the genetic alteration of microorganisms for military purposes was, in fact, already expressly prohibited before the Plasmid Working Group issued its urgent appeal and that this practice continues to be prohibited today. They point to a major international treaty, the 1972 Biological Weapons Convention, which has been signed by about half the nations of the world, including both the United States and the Soviet Union.

The treaty, which went into effect in 1975, seems explicit enough concerning biological weapons research. Its parties pledge that they will refrain from producing "microbial or other biological agents, or toxins, whatever their method of production, of types or in quantities that have no justification for prophylactic, protective, or other peaceful purposes." They also vow that they will "never in any circumstances, develop, produce, stockpile, or otherwise acquire or retain" biological weapons. Unlike any other modern international arms treaty, the Biological Weapons Convention even mandates that the nations disarm themselves — destroying any existing stores of toxic agents.

Because this landmark treaty was formulated before the appearance of recombinant DNA techniques, it could not possibly have anticipated the potential threat of genetically engineered biological weapons nor issued a specific condemnation of their creation. Nonetheless, a number of leading experts in biological weapons, including Matthew Meselson, an internationally respected Harvard molecular biologist, insist that the treaty's broad prohibitions against all biological weapons already provides a basis for adequate safeguards against the prospect of genetically engineered weapons.

Critics of the treaty, including Susan Wright, a science historian at the University of Michigan, and the noted molecular geneticist Robert Sinsheimer, have argued that a number of key phrases in the treaty could be interpreted in ways that might seriously weaken its ability to prevent an international biological arms race. In their view, the Biological Weapons Convention forbids, for example, the "development" of biological and toxic weapons — yet permits "research" on them. It outlaws the use of biological weapons for "hostile purposes or armed conflict" — yet allows parties to retain them in quantities suitable for defensive needs. And it provides no upper limit on "acceptable" stockpiles of biological weapons.

However, Meselson and other supporters of the biological weapons treaty insist that it remains our single best hope for continued global biological disarmament. During the past two years, that optimism has been borne out to some extent by efforts on the part of the United States, the Soviet Union and other parties to reduce the possibility that genetic engineering techniques will be used to forge new generations of biological weapons. These measures include the exchange of information concerning each nation's special, high-containment laboratory facilities where hazardous genetic engineering experiments are carried out; the reporting of unusual outbreaks of diseases that might be associated with biological weapons research; increased sharing of research findings concerning protection against biological warfare agents; and annual meetings by parties to the treaty. These and other developments suggest that our most effective response to the prospect of genetically enhanced biological weapons may be to work constructively with existing international agreements rather than to hastily discount them or to request additional funding for "defensive" biological weapons research programs.

A Hypothetical Biological Warfare Scenario

The prospect of new generations of genetically engineered biological weapons is indeed a legitimate cause for public concern. Unfortunately, the topic of biological warfare has not yet emerged as a compelling concern to most people. To most, it smacks of alarmist science fiction and has failed to arouse widespread public indignation, in large part because of the lack of reliable information on the subject.

But our discussion of the use of recombinant DNA techniques to build biological weapons would be incomplete without raising a number of disturbing personal questions that arise only when one contemplates the real horror of biological warfare in action. What would it be like to experience a secret, state-of-the-art biological warfare attack in the future — one, for example, that targeted civilian populations and possessed no clearly drawn battle lines or official declaration of war? How might the victims of a highly selective, clandestine assault suffer from the artificial epidemic? How might other members of society — those unaffected by the artificial epidemic and unaware, perhaps, of its military origins — respond to the human suffering they witnessed around them? And, in the end, what kind of world would we be creating by passively permitting individuals or nations not only to design novel biological weapons but also to deploy them against fellow human beings?

Because of the secrecy that has always limited public discussion of biological warfare, most of us would find it difficult to formulate informed responses to such vital questions. Yet, unless we search for at least tentative answers to them, our image of modern biological warfare is bound to remain abstract, intellectual and devoid of a visceral sense of the grim reality of this unconventional mode of warfare.

In the absence of documented historical precedents of large-scale biological attack or a major laboratory accident involving recombinant biological weapons, we are compelled to draw on fictional scenarios of biological warfare to try to come to terms with the possible magnitude and scope of the hostile use of living organisms.

Consider, for example, the following thought experiment. Imagine that a team of scientists is engaged in a secret biological weapons research project on behalf of a fictional, extremely fanatical military regime or an underground terrorist organization. Their goal: to use recombinant DNA techniques to create a deadly human virus that might eventually be released to selectively infect a specific minority group that is perceived as a threat to that organization's plans to maintain or achieve political power. Imagine, further, that the virus the scientists hope to genetically modify into a future ethnic weapon happens to be an infectious, RNA-containing retrovirus closely resembling the virus we know as the Human Immunodeficiency Virus (HIV), or AIDS virus — the agent responsible for the fatal immune disorder known as Acquired Immune Deficiency, or AIDS. Finally, imagine that during the course of this clandestine research, a scientist commits a terrible blunder and a quantity of the potent, AIDS-like pathogen is accidentally released long before it has been genetically honed into the highly selective, efficient ethnic weapon its creators had envisioned. That is, the unfinished AIDS-like virus inadvertently begins to infect adjacent populations before the scientists have confirmed its host selectivity or constructed vaccines to protect nontarget populations.

We emphasize that we are not suggesting that the AIDS virus itself had its origins in biological weapons research in some part of the world. It is true that there have been isolated, and to date utterly unfounded, media reports containing claims by a handful of scientists in the United States, Europe and the Soviet Union that the AIDS virus could have arisen from a biological weapons experiment gone awry. However, this notion remains nothing more than speculation and has obvious potential for abuse as superpower propaganda.

While the precise origin of the AIDS virus remains largely a mystery, the prevailing view among reputable scientists is that the AIDS virus arose naturally decades ago from a relatively harmless animal virus resident in populations of wild African green monkeys in equatorial Africa. It is thought that this ancestral virus could have been transmitted to humans through the exchange of blood from a scratch or a bite wound. During the transition from animal to human virus, scientists speculate, the virus was somehow transformed into a deadly, more virulent human pathogen.

Like the AIDS virus, our hypothetical, renegade biological warfare agent can be transmitted by virtually any behavior that results in the exchange of infected human body fluids, including blood, semen or vaginal secretions — a common occurrence in human sexual activities as well as in such nonsexual situations as the sharing of contaminated hypodermic needles or transfusions using unscreened blood or blood products. As a result, the AIDS-like virus spreads quickly through minority populations, which happen to be geographically concentrated in the region surrounding the clandestine biological weapons project. The virus thrives in these human hosts, who represent its intended target. But because the host range of the crudely modified virus has not yet been refined to pathogenic perfection, the microbe can also infect anyone who inadvertently exchanged body fluids with the first victims of the artificial epidemic — including, ironically, the scientists responsible for the accident. As the AIDS-like infection begins to leak into nontarget populations, it gradually ignites a global epidemic, or pandemic, of unknown origin that, in a matter of years, affects millions of people.

And like the AIDS virus, this manmade virus causes a new disease that is clinically unknown to physicians. It too wreaks its havoc on the human body primarily by destroying a specific subgroup of white blood cells in the immune system — called helper T-cells — that are critical control elements in the body's cellular response to infection. As a result, infection with this novel, genetically engineered virus cripples the immune defenses of most of its victims. They die not from the direct action of the virus but from an assortment of opportunistic infections — characteristic forms of cancer, parasitic diseases and other illnesses that are normally rebuffed by a healthy human immune system.

As our fictional AIDS-like biological weapon spreads, we might expect it to leave in its wake an epidemiological pattern resembling in some ways that of the real AIDS epidemic. First, we might expect a biological warfare agent designed to meet the goals of our fanatical political organization to be targeted at specific minority groups judged by these malevolent minds to be the genetically "least desirable" elements of society. If the organization's eugenic agenda were blatantly racist, it is conceivable that its targets might include some of the same groups that are today disproportionately affected by the AIDS global epidemic. Among them: black, heterosexual Africans and North American blacks, Hispanics, homosexuals and intravenous drug abusers. Second, we might expect that agent to kill its victims with a relentless fury — of a kind rarely encountered in most naturally occurring diseases. Third, we might expect the agent to sidestep not only the human body's elegant natural defense system but also existing defensive measures of modern medicine — ranging from drug treatments to preventive vaccines. Fourth, in the absence of such immunological and medical controls, we might expect very rapid infection — triggering an epidemic that however horrifying to its victims and their loved ones, might be viewed by some sectors of society as ideologically convenient. Fifth, we might expect the sudden appearance of a biological warfare agent with such unconventional characteristics to seem to be a bona fide medical mystery — at least for a time — to virtually everyone except the scientists who were responsible for it. And finally — in light of the inevitable uncertainties of designing diseases and the capacity of biological organisms to change and adapt to their new surroundings — we might expect to see the pathogen gradually diffuse beyond the intended "boundaries" of those populations originally targeted for infection. In a dramatic display of the impossibility of human control over the forces of disease, it might begin to spread rapidly in human populations characterized by other racial origins or sexual habits, including members of the very group whose research scientists had first conceived of the malevolent virus.

Our point in using an imaginary AIDS-like biological agent as an illustration of a misapplication of molecular genetics is simply to suggest the terrifying extremes to which genetic knowledge could be exploited in service, for example, of political or religious extremists, or, for that matter, of emotionally disturbed individuals with scientific expertise. The point is to remind us how narrow the boundary is between the benign microorganisms with which we share our daily lives and the rare pathogen that can shatter our health. This thought experiment is meant to suggest that if we, as a society, willingly accept the application of genetics for building weapons, we had better prepare ourselves for a nightmarish world. For by harnessing the forces of disease to settle human conflicts, we will be entering into a Faustian bargain that could destroy the very qualities — compassion, empathy, altruism — that the majority of us most cherish in human beings.

Toward Scientific Responsibility

Throughout history, knowledge has presented us with a double-edged sword. Human innovation — from the invention of knives, spears and arrows to the design of intercontinental missiles armed with nuclear warheads — has simultaneously advanced both our means of survival and our capacity to harm one another. Yet we are not completely powerless to limit the growing militarization of science.

Strong, verifiable international treaties dedicated to preventing the use of modern genetic technologies to build biological weapons can serve as a major deterrent. We can also begin to demand of our life scientists the same sort of moral accountability — in the form of a universal code of professional ethics — that we have expected of our healers. Today there exists no compelling set of moral principles to guide new generations of genetic engineers through the difficult personal and professional decisions many of them will be forced to make. Nor have universities granted them the opportunity during their critical years of academic training to adequately search for moral principles of their own.

It would not be easy to reach consensus on the social responsibilities of scientists. But, in the end, such a code might do much to reaffirm the values that many scientists today consider most imperiled by a growing public sentiment that society has a right to place limits on the future applications of genetics. The most important of these would be the enduring value — despite the inevitable risks — of pursuing scientific knowledge for its own sake. And, at the very least, such a code would have to establish barriers to scientific research dedicated to degrading or destroying human life.

It is also essential that such a code contain provisions that would eliminate the shrouds of secrecy that have prevented open and informed public debate concerning biological warfare issues. There is no way of knowing the precise nature or scope of secret biological weapons research that has been carried out over the years in the United States, the Soviet Union and other countries. But it is safe to say that the mere existence of such classified scientific research has helped to distort public perception of biological warfare. In biological weapons research, secrecy can assist a nation in achieving a technological advantage over its opponents and also offer genuine propaganda value. But secrecy can also result in exaggerated public fears. It can inadvertently lead to the discrediting of legitimate scientific research projects that only appear to have military usefulness. And it can contribute to an international climate of hostility and distrust and thus further legitimize or even unnecessarily escalate existing biological weapons programs.

Society's surest protection against the misuse of scientific knowledge is — despite the obvious risks — mandatory publication of all research findings. If the scientific community adopted a professional code of ethics that explicitly forbade classified research, public vigilance might drastically limit the opportunities for individual scientists using modern genetics to intentionally harm fellow human beings.

For, in the long run, secrecy is the enemy of legitimate science. It paralyzes the free exchange of information and unfinished ideas that leads the scientific community slowly toward provisional scientific truths. Secrecy encourages scientific elitism. And it denies the public a decisive voice in determining the priorities of the scientific research that it directly or indirectly pays for.

CHAPTER 10

ENVIRONMENTAL DAMAGE TO DNA: DEVELOPING A SENSITIVITY TO THE SUFFERINGS OF GENES

> **GENETHIC PRINCIPLE**
> The information contained in genetic molecules is vulnerable to loss through mutations caused by sunlight, radioactivity, chemicals and other external mutagenic forces. Each of us has a responsibility to develop an awareness of potential mutagens in our immediate surroundings and to seek to minimize environmentally induced damage to our DNA.

> *"Why in the world did He ever create pain?"*
> *"Pain?" Lieutenant Scheisskopf's wife pounced upon the word victoriously. "Pain is a useful symptom. Pain is a warning to us of bodily dangers."*
> *"And who created the dangers?" Yossarian demanded. ". . . Why couldn't He have used a doorbell instead to notify us, or one of his celestial choirs? Or a system of blue-and-red neon tubes right in the middle of each person's forehead. . . ."*
> — Joseph Heller, *Catch 22*

The Nature of Genetic Mutation
Despite the breathtaking precision with which the coded messages of DNA molecules are transcribed into hereditary replicas or translated into biologically useful products in the cell, mistakes

238

— however rare — inevitably occur. Errors that affect a cell's genetic memory — anything from a harmless substitution of a single nucleotide base to the gain or loss of an entire chromosome — are referred to as *mutations*.

Errors in genes are generally less likely to arise from flaws in the cellular machinery of biological inheritance than from the endless storm of external forces — chemical, physical or biological — that rages continuously around them, always threatening to disrupt the graceful dances of genes. We call these environmental sources of genetic chaos *mutagens*. And they can assume an astonishing variety of forms.

They can be chemical substances — ranging from natural biological poisons secreted by plants and fungi to synthetic food additives and genetically toxic industrial chemicals. They can be natural forms of energy. The same invisible shaft of short-wavelength ultraviolet radiation from the sun that tans your skin at the beach, for example, can be mutagenic. So can searing bursts of *ionizing radiation* — the subatomic rubble of protons, electrons and gamma rays that is ejected at the speed of light from radioactive materials. Even living organisms — certain viruses, for example — can leave behind a trail of minute alterations in the DNA sequences of host-cell chromosomes during the cycle of infection. Finally, random alterations are also etched into genes by seemingly phantom mutagenic forces — ones that cannot be reliably traced to their source. Geneticists refer to these collectively as "spontaneous" mutations. But their spontaneity arises not from any intervention of mystical forces in nature but from molecular mechanisms taking place in the genetic apparatus to which, for the moment, we remain blind.

Environmental damage to a gene is seldom a dramatic occurrence. In principle, a mutation can arise in the human genome from an event as mundane as munching on a peanut butter sandwich. In this case, the culprit is a mild chemical mutagen — one of thousands of natural or synthetic compounds that can cause lasting damage to the structure and function of genes. Known as aflatoxin, it is excreted by colonies of a microscopic mold —

members of the genus *Aspergillus* — that sometimes contaminate peanuts or peanut products. Even in quantities too minute to induce clinical symptoms of food poisoning, this unseen fungal toxin can trigger a tiny but measurable increase in mutations — though not of a magnitude that should cause one to relinquish forever the gustatory pleasures of eating peanut butter.

Even sipping a freshly brewed cup of coffee at breakfast can, theoretically, result in a minute increase in mutations. The active ingredient in coffee is caffeine — roughly 100 milligrams in every cup — which exerts a variety of pharmacological effects on the human body. Among other things, caffeine excites brain neurons, accelerates the heart rate and stimulates the kidneys to excrete more urine — all familiar symptoms to most habitual coffee drinkers. In some organisms, caffeine is known to cause subtle damage to DNA, and for years many scientists suspected it was also mutagenic in humans and other mammals.

Recent research suggests that the human body is normally equipped to detoxify caffeine before it has an opportunity to disturb the information harbored by human DNA sequences. Nonetheless, stale coffee that has been allowed to brew for several hours can form highly reactive chemical compounds known as peroxides that act as weak mutagens. Again, the toxic effects of these substances on human DNA are likely to be so minuscule compared with those of all sorts of other mutagenic forces around us that they should not be viewed with alarm. But their impact can be monitored in the DNA molecules of bacteria and other laboratory microorganisms in carefully controlled laboratory experiments.

Cellular Systems for DNA Repair

Mutations are nothing new to genes. From the instant the first self-replicating hereditary molecules appeared, adrift in the earth's shallow primeval seas, they have been under siege by mutagenic agents of one kind or another. A thin atmospheric layer of ozone gas — a faintly bluish, malodorous, triatomic form of oxygen generated by the photochemical action of solar radiation striking oxygen that has been released by photosynthetic green plants over periods of millions of years — now shields DNA molecules from excessive quantities of ultraviolet (UV) sunlight. Water is also a barrier to UV. But primitive life forms attempting to creep out of the seas and lakes enjoyed no such protection. As a result, the genes of early life forms suffered grievous damage from exposure to the mutagenic rays of unfiltered solar radiation. Then, as now, genes were pelted by a steady rain of ionizing radiation from naturally occurring elements in the surrounding soils, seas and skies. And their messages were continuously garbled by naturally occurring environmental substances that seeped into cells and caused mutations.

In such hostile surroundings, genes, by necessity, gradually evolved biological mechanisms to defend themselves against the forces of mutation. While natural selection would favor life forms that could cope with the hereditary havoc wreaked by mutagens, it would do so only up to a point. As we have discussed in previous chapters, mutations — as the ultimate source of the raw genetic variability upon which natural selection operates — are a fundamental ingredient in the evolutionary process. So one would not expect genes ever to become so efficient at policing mutational damage to DNA as to be able to erase them completely.

Nonetheless, organisms have over time evolved a number of ingenious, genetically controlled mechanisms that serve to minimize random loss of the hard-won hereditary information etched in ordered base-pair sequences of DNA. In our technological age, when rising levels of mutagenic agents in the environment are part of the price of material progress, the mere existence of such ancient biological strategies for DNA repair should offer us at least some small comfort. For they remind us that genes were battling mutagens long before industrialized societies arrived on the evolutionary scene. Yet the question remains: At what point might the modern quota of environmental mutagens — physical, chemical and biological — so shift the delicate equilibrium between gene mutation and repair that it threatens our survival and that of fellow species?

UV Damage and Repair

One of the early challenges faced by primitive life forms was to minimize mutational changes to genes caused by solar radiation. When rays of ultraviolet light strike a DNA molecule they leave behind a characteristic chemical spoor. As the rays pierce a living cell and strike a molecule of DNA, they can create all sorts of subtle chemical havoc in the double helix.

The most important of these occur when the rays transfer a portion of their energy load to the molecular subunits of DNA, creating new chemical bonds that result in kinks, bulges and other physical contortions in the symmetrical architecture of DNA. Since DNA replication and transcription are critically dependent on the precise physical alignment of complementary nucleotide bases on opposite strands, these injuries can lead to inaccurate base pairing and therefore to errors in the genetic messages coded in replicas of a sun-scarred DNA strand.

The most typical mutation created by UV light passing through DNA involves neighboring pyrmidine molecules — the thymine (T) and cytosine (C) nucleotides, which form the steps in the DNA spiral staircase. The energized thymines bond together, creating a distinctive tandem molecule, or dimer, along one strand called a thymine-thymine "bridge." During DNA replication, this swollen dimer so distorts the match of complementary bases that its two thymine bases can end up mistakenly matching with a guanine (G) molecule, instead of the usual adenine (A) prescribed by DNA's base-pairing rules. Because each such mismarriage between two bases represents a flaw in a gene's original message that can be passed on to progeny cells, it constitutes a genetic mutation.

To counter this sort of damage from UV light, cells are equipped with special maintenance teams of genetically encoded repair enzymes. One — found in the cells of most bacteria, as well as in those of many higher organisms, is spurred into action only if a cell is first illuminated by visible sunlight (see Figure 10.1.). Energized by solar rays, an enzyme quickly homes in on the chemical wound in the DNA molecule caused by UV rays.

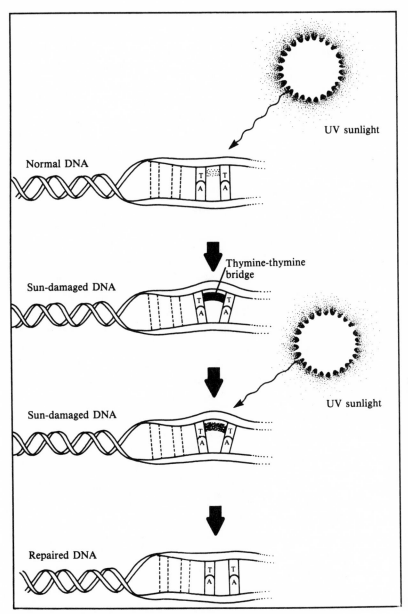

FIGURE 10.1: UV Damage and Repair Mechanism

Then it splits the bulky dimer bridge, freeing the resilient double helix to spring back into its original symmetrical shape.

In mammalian species like ourselves, oddly enough, normal DNA replication processes automatically pair thymine dimers with the appropriate adenine bases and are therefore not disturbed by a thymine-thymine wound. But if the solar lesion turns out to be a cytosine-cytosine dimer, leading to a mismatch with adenine on the opposite strand, corresponding UV repair mechanisms swing into action to excise the offending dimer bridge.

Other DNA Repair Mechanisms

UV-induced damage to DNA can manifest itself in a variety of other ways. The solar wound can take the form of missing or modified nucleotide bases, a clean break in one or both strands of the double helix, or chemical bonds that establish disruptive crosslinks between complementary DNA strands. But mutational damage is not limited to these. Virtually any structural change in the double helix that either alters the coded informational content of a gene's message or disturbs the intimate base-pairing mechanisms at work in DNA replication, protein synthesis and chromosomal crossing over qualifies as mutational damage. As a result, cells have evolved alternative genetic repair systems capable of coping with these mutational setbacks as they arise.

Among the most important of them is a generalized repair mechanism known as the excision-repair system (see Figure 10.2). It consists of a highly coordinated assortment of enzymes that pursue a slightly different strategy than the enzymes of the more specialized UV-repair system. In addition, this enzyme team carries out its maintenance work covertly — strictly under the cover of darkness. The basic plan is to surgically remove an offending DNA sequence that contains a mutational error in one strand of DNA and then to replace it with a freshly synthesized error-free sequence.

Working together as a unit, several excision-repair enzymes spot the mutation site, clamp on to the lesion and neatly snip

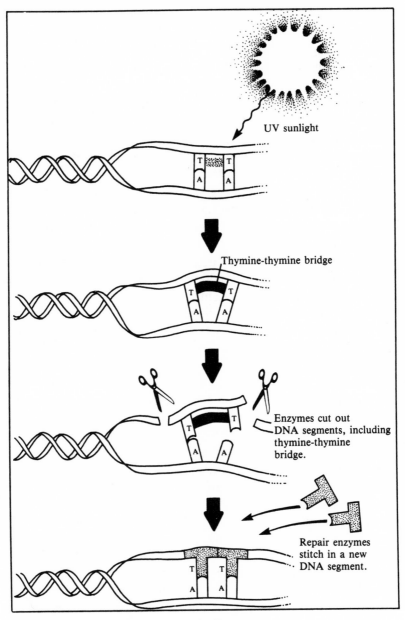

UV sunlight

Thymine-thymine bridge

Enzymes cut out
DNA segments, including
thymine-thymine
bridge.

Repair enzymes
stitch in a new
DNA segment.

FIGURE 10.2: Excision-Repair System

out a segment of DNA approximately 20 bases in length that harbors the mutation. Then additional enzymes appear on the scene to stitch a corrected version of the missing section methodically into place. To carry out this impressive feat, they use the complementary DNA strand opposite the gap as a sort of mirror-image molecular memory of the amputated sequence. After they have matched each exposed base on the template strand with a complementary mate, enzymes also weld the freshly synthesized stretch of DNA into the spiral scaffolding of the double helix, just as they do during the genetic dance of DNA replication.

Like the others, the excision-repair system is so seamlessly woven into the metabolic fabric of a living cell that its often heroic DNA-salvaging efforts can easily go unnoticed. Often it is only when the repair system suffers a catastrophic collapse that we are forced to take notice of its pivotal role in stemming the loss of genetic information to random mutation. This is precisely what takes place in a rare inherited human disease known as *xeroderma pigmentosum*. The illness can be traced to a recessive gene that is found in homozygous double dose in victims of the illness. The defective gene cripples the intricate excision-repair mechanism by failing to code for adequate quantities of one of the critical members of the enzyme team. This loss renders cells of the epidermis unusually vulnerable to damage from UV light, even in the relatively low dosages found in sunlight.

Gradually overwhelmed by UV damage to their DNA that they are no longer equipped to repair, skin cells careen into crazed, precancerous cell divisions that grow increasingly frenzied. In time, as the number of mutations in the cells' sunstruck DNA increases, clusters of epidermal cells begin to proliferate into visible lesions in the victims of *xeroderma pigmentosum*. First the renegade cells pile up in darkly pigmented mounds that look like freckles. Then they develop into tumorous swellings that disfigure exposed skin. Finally, they may develop into full-blown skin cancers that can even become life threatening. These cancerous growths stand as graphic testimony to the power of mutations — in the absence of built-in biochemical forces of repair — to disrupt the flow of genetic information so essential to normal cellular growth and reproduction.

Cells depend on an assortment of other schemes to deal with genetic errors as they arise. Some of the repair mechanisms, for instance, pinpoint a localized mutation site and neatly remove it without affecting neighboring DNA sequences. Others perform bolder genetic surgery — snipping out DNA segments extending several nucleotides on either side of a mutational wound and then inserting a newly synthesized replacement.

One system, known as postreplication repair, even dispatches enzymes to patrol the double helix in search of minor flaws in the steps of DNA's spiral staircase that have arisen from improper pairing of bases during DNA replication. Whenever a mismarriage of bases is detected in a newly replicated double-stranded molecule, the system swings into action. First, enzymes must determine which of the two strands in the double helix harbors the incorrectly matched base. Their technique is ingenious. Following faint chemical clues, they are able to distinguish between the original template DNA strand and the newly minted one. Since the template sequence bears the correct genetic message, repair work is directed to the garbled complementary one. Just as in excision repair, enzymes remove a length of the strand that encompasses the mismatched base. Then, using the opposite strand as a template once again, they stitch the proper nucleotide-base mate into place, effectively erasing the original random replication error from the genome.

Nature's Balance between Mutation and Repair

The net effect of such gene-maintenance systems is to establish a dynamic equilibrium between the opposing processes of genetic mutation and repair. Inevitably some mutations go undetected. These rare, one-in-a-million errors combine to form the burden of mutations that is borne by the gene pools of every living species. As we have noted in previous discussions of natural selection, each such mutation does offer a small possibility of improving a gene's time-tested message. But the vast majority of unrepaired mutations is likely to be injurious — with mild to lethal effects. Once they are translated into defective enzymes or structural proteins, they are far more likely to become obstacles than aids to the clockwork efficiency of cellular metabolic processes. As a result, new mutations often result in the death of a cell. In complex, many-celled organisms, the effects of mutation may be more enigmatic — triggering curious birth defects, time-delayed cancers and other malfunctions within the larger community of interacting cells. Only the most exceptional mutations prove to be evolutionary assets. By coding for a timely innovation in a gene product or process, these few fortuitous errors add to an organism's prospects of surviving and reproducing.

Mutation and Disease

Part of the price we must pay for the dependence of natural selection on a continuous supply of new mutations is an endless toll of often devastating human genetic diseases. Chromosomal mutations — imperfections in the distribution of chromosomes to cells — are responsible for a variety of relatively rare genetic illnesses. The best known of those is Down's syndrome — a disease marked by the presence of three copies of human chromosome number 21 instead of the usual two — and the most frequent definable cause of severe mental deficiency in the United States. Together, these abnormalities in chromosome structure or number affect approximately 1 out every 200 newborn infants. Furthermore, most of them appear to represent recent genetic errors — arising anew each generation in the reproductive cells of the parents of a victim, rather than drifting intact through a long lineage of ancestors. As such, they are a reminder of the utter impossibility of ever fully "cleansing" the human gene pool of genetic ills.

Mutations can also trigger disease at the level of individual genes. A single-gene, or monogenic, disease can be transmitted in the form of a dominant Mendelian allele, initiating the symptoms of the illness in those who inherit at least one copy of the mutant gene from their parents. Single-gene mutations may also be transmitted as recessive alleles that lie dormant and unexpressed except in genomes that have received a double dose — one from each parent — of the pathogenic gene. Still other single-gene mutations are inherited without either clear-cut recessive or dominance patterns.

Most of the more than 3,000 genetic diseases that can be traced to a defect in a single gene occur so infrequently in the human population that their names have an exotic ring to most of us — bilateral retinoblastoma, phenylketonuria and galactosemia, to name just a few. But together these monogenic diseases are thought to be responsible for more illnesses — from rare hereditary cancers to inherited blood diseases and deadly enzyme deficiencies — than are the more visible chromosomal mutations.

Mutations must also affect human health in ways that remain invisible to us. Mutant genes undoubtedly play a role in the occurrence of a number of more familiar diseases such as certain forms of epilepsy, allergies, mental illnesses and diabetes. But we do not yet fully understand the relative contribution of heredity and environmental forces in the pathology of such polygenic diseases — nor even the exact number or identity of the critical causative genes. For this reason, we can only guess how often new mutations might be involved in triggering such diseases. In addition, mutational damage to DNA molecules can result in some forms of cancer. If a mutation arises, for example, in the midst of a proto-oncogene — a highly vulnerable DNA sequence coding for an enzyme or a regulatory protein vital to normal cell growth — it can instantly transform healthy mitotic cell divisions into uncontrolled cancerous divisions. Finally, by using recombinant DNA techniques, mutations that are so minor they do not even visibly surface in the phenotype of an organism can now be routinely spotted in human DNA. Called restriction-enzyme fragment-length polymorphisms, this category of minute mutational blemishes in the double helix includes mutations that are so subtle that they simply cannot be detected in any other way than by directly probing genetic molecules with restriction-enzyme scalpels.

Disrupting the Genetic Balance

Up to this point, we have avoided placing any final judgment on mutations. In the natural world, mutations must be seen as an ethically neutral phenomenon — one measure of the imperfections of heredity that is as old as the origins of life itself. For any species to survive, there must be a balance between the genetic costs and benefits of mutation. The costs can be defined as the erosion of genetic information, and the benefits as increased genetic variability. But by introducing significantly increased levels of genetic toxins into the human environment, modern industrial societies can shift the balance as unacceptably high levels of DNA damage threaten an increase in genetic abnormalities, disease and death.

UV Radiation

A graphic case of the effect of human activity on genetic mutation is the impact of air pollution on the ozone layer. In the 1970s, scientific reports that synthetic chemicals such as fluorocarbons, which are found in commercial refrigerator coolants, aerosol products and other consumer products, appeared to be eroding the ozone layer led to a public outcry and international efforts to ban fluorocarbons from the marketplace. Now, a decade later, new studies suggest that the deterioration of the fragile ozone shield by air pollutants continues unabated.

Beginning in 1979, scientists have documented a thinning in the ozone layer approximately 20 kilometers above the Antarctic continent. This tear in the ozone veil that envelops the planet seems to expand each year. If it continues to grow, it will eventually begin to bathe towns located in southern Argentina, and beyond, with higher doses of UV radiation. This, in turn, can be expected to result in higher mutation rates in the skin cells of local residents, resulting in a corresponding rise in the incidence of skin cancers. The possible ecological consequences of a rising influx of UV radiation on critical components of ecosystems around the world remains unknown.

Ionizing Radiation

Nuclear technology may offer the most dramatic example of damage to DNA resulting from human activity. Our midcentury mastery of atomic fission and fusion has allowed us to tap a vast reserve of energy and has transformed global geopolitics. At the same time, it has exacted a genetic price, which went undetected for years, in the form of increased levels of mutation and cancers resulting from exposure to radioactive wastes.

The radioactive debris of nuclear reactions — including X-rays, gamma rays, alpha rays, beta rays, electrons, neutrons, protons and other high-energy waves and particles — is mutagenic. Collectively, it is called ionizing radiation — in contrast to such nonionizing radiation as sunlight — because of its ability to strip electrons from, or ionize, atoms in its path. Ionizing radiation varies in its ability to penetrate the dense water-laden tissues of the human body. Beta rays, for example, consist of streams of energetic electron particles that quickly exhaust themselves in the outermost layer of skin cells. Gamma rays, in contrast, are a more powerful form of ionizing radiation, resembling medical X-rays, except that they originate in the nucleus of an atom rather than in the clouds of electrons that swirl about the nucleus. Gamma rays can penetrate more deeply into tissues, leaving behind a longer trail of cellular damage.

The sudden transfer of energy from ionizing radiation to the atoms making up DNA, RNA, proteins and other important biological molecules can drastically alter a living cell (see Figure 10.3). Ionizing radiation can bombard DNA directly as it plunges into a cell, dislodging electrons from individual atoms in the double helix or sometimes even scoring direct hits that sever one or both strands of the double helix. Radiation can also injure DNA indirectly by leaving in its wake a froth of electrically unbalanced ions, electrons and other assorted charged particles. These often collide with other molecules adrift in the cytoplasm that neutralize their charge, rendering them less reactive. If the collision happens to involve a molecule of DNA, the result can be direct damage to a gene's chemically coded message — a genetic mutation. Ions can also collide with oxygen, water and other nonhereditary molecules circulating in the cytoplasm. These collisions can create highly reactive intermediate substances capable of inflicting secondary damage to DNA, often long after the original source of ionizing radiation has been removed.

Our genes face continuous bombardment from so-called background radiation — ranging from naturally occurring radioactive elements such as uranium and thorium in the soil and radioactive potassium, phosphorus and calcium in our bodies to cosmic rays from outer space. In fact, the only isotope of the vital element potassium available to living organisms happens to be radioactive. So life has continuously been compelled to make compromises with the mutagenic forces of ionizing radiation. But genes now also must contend with radiation from a variety of human activities.

In most modern industrialized nations, the single most important new source of ionizing radiation affecting human genes is medical and dental X-ray examinations. But, depending on one's geographic location, health and lifestyle, there may be a number of others. Among them: radioactive wastes from the nuclear power industry, radiation treatments for cancer, radioactive tracer materials injected into the body for medical diagnosis, and occupational hazards involving the handling of radioactive materials.

Added to this are such chronic sources as radioactive fallout from decades-old atmospheric nuclear weapons tests. Trapped in the upper reaches of the atmosphere as a result of experimental detonations during the 1950s, this reservoir of airborne radioactive debris continues to cascade slowly into lower atmospheric layers, eventually settling on the surface of the earth. Humans, along with many other species in nature, are highly vulnerable to specific radioactive *isotopes*, or radioisotopes — unstable elements that differ from related nonradioactive elements in the number of neutrons they carry in their nuclei — that can readily enter the body through contaminated air, water or food.

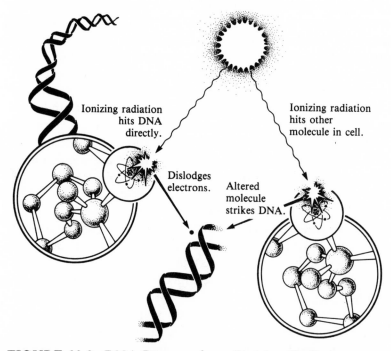

FIGURE 10.3: DNA Damage from Ionizing Radiation

If the radioactive isotope strontium 90, for instance, dusts a dairy pasture and is consumed by grazing cattle, it tends to follow a bovine metabolic path paralleling that of the nonradioactive element calcium. As a result, strontium 90 is biologically concentrated in calcium-rich cow's milk. When humans drink the contaminated milk, strontium 90, again masquerading as nutritious calcium, accumulates in living bone tissue, where it can bombard cells at close range with damaging ionizing radiation for decades. After entering the human food chain, other radioisotopes suffer distinctively different fates. Once eaten or inhaled, relatively short-lived radioactive iodine 131 atoms, for instance, are concentrated in the human thyroid gland; cesium 137 atoms, a longer-lived radiation source, follow other metabolic routes.

The nuclear accident that took place at the Chernobyl nuclear power plant in the Soviet Union on April 26, 1986 — the worst in recorded history — is a recent reminder of the ability of airborne radioactive wastes to contaminate not only human beings but entire ecosystems. As high-altitude winds carried nuclear debris vented by the Chernobyl accident over northern Europe, clouds of radioactive particles settled on vast areas of tundra — the habitat of large herds of semidomesticated reindeer raised by Laplanders. The tundra is covered with lichen — a rootless carpet of vegetation that is a symbiotic mix of algal and fungal organisms — which absorbed radioactive moisture from the atmosphere.

Reindeer ingested the contaminated lichen that serves as their traditional winter fodder, and radioisotopes such as cesium 137 and cesium 134 became concentrated in their body tissues. The reindeer were quickly perceived as a potential health risk to human populations that consume reindeer meat — although controversy remains as to whether the relatively low levels of radioactivity found in the contaminated carcasses will add significantly to existing spontaneous mutation rates in humans and cause a noticeable increase in the incidence of cancer in the decades ahead.

Nonetheless, in 1987 alone, tens of thousands of slaughtered reindeer — roughly half of the annual harvest — are expected to be declared unfit for human consumption. Because it takes about 30 years for the radioactivity of cesium atoms incorporated into living reindeer muscle to decline by one-half, this new source of ionizing radiation in the tundra ecosystem can be expected to pose lingering health concerns.

It would be difficult to imagine a more dramatic challenge to the integrity of DNA molecules in the gene pools of human and nonhuman species than a full-scale nuclear war. Geneticists have had only limited experience with the effects of massive doses of neutrons and gamma rays on human health. But they do know that mutational damage caused by ionizing radiation generally increases in direct proportion to the level of exposure to radioactivity and that there appears to be no absolute threshold of radiation dosage below which biological damage does not occur.

Unlike the controlled beams of chest X-rays, for example, which are intended to affect body cells only, ionizing radiation from a nuclear blast would affect the genes of both somatic, or body, cells, and germ-line, or reproductive, cells. Thus, in addition to triggering the expected somatic-cell symptoms of lethal and semilethal mutations, cancers and radiation deaths in survivors of the blast, ionizing radiation has the potential to also trigger mutations in the genetic molecules of germ cells, some of which might not be visibly expressed for generations. Furthermore, by subjecting the gene pools of many species to extraordinary levels of mutagenic radiation, nuclear explosions and their long-term legacy of radioactive fallout could wreak genetic havoc on nonhuman species playing crucial roles in ecosystems around the world.

The survivors of the bombings of Hiroshima and Nagasaki in World War II offer a unique preview of the possible mutagenic risks of global nuclear war. These victims and their descendants have been studied intensively for decades now, in an effort to determine the genetic cost of such exposure. It is clear from these investigations that DNA in the somatic cells of some survivors did suffer immediate damage. But it also appears that ionizing radiation is much more efficient at simply killing cells by disrupting their intricate metabolic processes than it is at inducing genetic mutations in them.

Many cases of acute radiation sickness and deaths following the explosions can be attributed in large measure to the disruption of mitotic divisions in cells of blood-forming bone marrow, digestive-tract lining and other rapidly dividing body tissues. Similarly, the tenfold increase in leukemia in victims during the 1960s, two decades after the bombings, demonstrates the delayed effects of DNA damage on somatic cells of bone marrow and other tissues. In this case, the surge in leukemia cases was a transient one, returning again to normal levels in following years.

Chromosomal abnormalities that can be detected in karyotypes, caused by the radiation-induced breakage and imperfect reunion of chromosomes, have often been used as a crude measure of DNA damage from nuclear weapons. Surprisingly enough, offspring of original Hiroshima and Nagasaki survivors have to date failed to show any increase in inherited chromosomal abnormalities. In fact, the frequency of chromosomal anomalies turned out to be slightly lower than that of nonirradiated control populations.

Nonetheless, such findings do not totally eliminate the possibility that germ-line cells of some bomb victims may have suffered more subtle forms of radiation-induced genetic damage that have simply not been detected yet. First, geneticists cannot yet catalog every sort of potential genetic abnormality present in a cell. Second, many radiation-induced mutations, as we have mentioned, may be lying latent in egg and sperm cells of these first progeny as either severe or mildly deleterious recessive traits that may not surface for generations. And, finally, the sudden, massive injection of nuclear radiation and contaminating radioactive debris into global ecosystems following a full-scale nuclear war could prove far more damaging to genetic molecules — human and nonhuman — than the comparatively minor and geographically limited mutagenic events that took place at Hiroshima and Nagasaki.

Even if solid scientific proof of the long-term genetic effects of nuclear warfare on human populations is still lacking, it is safe to assume that DNA in human germ cells is likely to be no less vulnerable to damage from ionizing radiation than is DNA in human somatic cells that are already known to suffer genetic damage from radiation.

Chemical Mutagens
If nuclear radiation is one of the most highly publicized threats to the informational integrity of DNA molecules, chemical mutagens may prove to be the most complex and insidious. Modern industrial societies have already injected more than 50,000 commercial chemical compounds into the environment. And, according to recent estimates, more than 1,000 new ones are introduced into the marketplace every year. Hundreds of these are

already known to be mutagenic, causing mutations as they interact with the DNA molecules of test organisms. About 90 percent of these mutagens are clearly carcinogenic, or capable of causing cancer. During our lifetimes, each of us can expect to be assaulted by a staggering array of possible genetic hazards — from medical drugs and cigarette smoke to household insecticides and food additives. Since the early 1940s, geneticists have known how to intentionally cause damage to DNA using a handful of chemical mutagens. But so rapidly has the inventory of environmental mutagens grown since then that today only an expert in genetic toxicology could hope to keep up with the latest scientific findings about the chemical hazards faced by genes.

Chemicals can alter DNA in a number of very specific ways. Some molecules collide directly with individual bases within DNA sequences, damaging them to the extent that they can no longer properly pair with complementary mates. Others play a game of chemical subterfuge. Because of their close structural similarity to one or another of the four nucleotide bases, they can slip in to replace a bona fide base in the spiral staircase of the double helix. But they later prove to be flawed impostors. During replication they do not reliably pair with the required base and so cause single-base errors, called *point mutations*, in the newly synthesized complementary DNA strand. Still others disrupt the double helix either by attaching themselves to a base or by inserting themselves between neighboring bases in a DNA sequence. These intrusions can distort the chemical properties of bases so that they can no longer abide by the strict base-pairing rules necessary for storing and transferring genetic information.

A complete catalog of known or suspected chemical mutagens is unavailable; the number of known and suspected substances is large and constantly growing. But even a partial list would include many technical names that are meaningless to most of us — vinyl chloride, ethylene oxide, dimethylsulfate and nitrogen mustard, to name just a few. It would also include some substances that are more familiar to most consumers. Among them:

- certain pesticides, including fungicides such as captan and a number of insecticides that are widely used, particularly in developing nations
- certain industrial chemicals, including some of the ingredients used in the plastics, textile and petroleum industries
- certain food additives, including nitrite, used as preservatives, and artificial sweeteners such as saccharin
- certain pharmaceutical products, including specific anti-tumor medications and antibiotic drugs

Our point in mentioning such examples of proven or potential mutagens is not to generate an exaggerated concern about possible chemical threats to human DNA but, instead, to underscore the incredible diversity of mutagenic agents in our daily environment. Mutagens can often be found in our homes, in our workplaces, in the air we breathe, in the water we drink, in the food we eat, even, paradoxically, in our medical prescriptions — although often in concentrations that are unlikely to cause harm. Living, as we do, in a veritable sea of synthetic substances and products — a number of which have been shown to be mutagenic — how can one hope to monitor all of the real or perceived genetic risks of chemical mutagens without becoming overwhelmed by the enormity of the task?

Tests for Mutagens
For the moment, our best hope lies in the judicious use of emerging laboratory techniques designed to predict the mutagenic effects of new substances. The best known of these is the Ames test, named for Bruce Ames, the University of California biochemist who developed it during the 1970s. The Ames test relies on cultures of genetically modified bacteria — members of the species *Salmonella typhimurium* — to provide, as we shall see, a greatly simplified laboratory model of the possible mutagenic effects of chemical substances on human DNA. The bacterial cultures are exposed to suspected mutagens under carefully controlled conditions, and the damage they inflict on bacterial DNA sequences

is measured. These findings are compared with those in unexposed "control" cultures, where DNA molecules are altered by spontaneous mutation alone. The results are then extrapolated to approximate actual human conditions.

The trick is to select bacterial targets in which new mutations can easily be detected and scored. To accomplish this, bacteria are equipped with a mutant gene that leaves them incapable of randomly synthesizing a vital amino acid nutrient called histidine. As a result, the cells can only grow on a laboratory food medium laced with supplemental histidine. When the bacteria culture is bathed in a solution that contains a chemical mutagen, a small fraction of the defective genes can be expected — strictly on the basis of statistical chance — to mutate back to the original DNA base sequence that originally coded for the critical enzyme in histidine synthesis. In this way, some colonies of bacteria are freed from their original histidine dependency and thrive on a histidine-free medium that formerly would have proved lethal to them. By counting the number of the *Salmonella* colonies that survive on the histidine-free medium, one can come up with a crude index of a chemical compound's mutagenicity — its capacity to directly interact with and disfigure DNA.

To further highlight new mutations, the genes that controlled the DNA excision-repair system in the bacteria are altered in order to disable it. As a result, any new mutations that are induced in the *Salmonella* test cultures are deliberately preserved, rather than being routinely erased from bacterial genomes like footprints on wave-washed sand. In addition, genes are altered so that bacteria are no longer capable of manufacturing their own protective protein coats. Without these genes, mutant bacteria will not be toxic to humans who might accidentally ingest the microbes. Finally, steps are taken to ensure that the Ames test cultures mimic metabolic processes in the human body that can transform harmless chemical substances into mutagenic by-products. Since much of this chemical modification is known to take place in the human liver, enzymes extracted from rat livers are added to the bacterial brew.

Even with these refinements, bacteria cannot be expected to mirror every conceivable effect a chemical mutagen might have on human DNA. In the first place, human DNA comes in distinctive chromosomal packages that differ from the structure and organization of bacterial gene-bearing bodies. Second, chemicals are modified in their passage through the body in ways that cannot be adequately mimicked by simple, liver enzyme systems. Third, we are long-lived multicellular creatures, equipped with genomes harboring tens of thousands of genes. While each of our body cells contains a complete diploid set of genes, only a small portion of them — those coding for the characteristic protein products of a specific type of cell — is likely to be active in a given tissue or at a given time. For this reason, mutational damage to DNA can lie buried in the human genome in a physiologically dormant state — as unexpressed genes or recessive alleles. Or, mutational damage can modify subtle metabolic pathways that have yet to be mapped by biochemists. It can play a role in initiating cancers that may take years to surface, and we may never be able to trace them to specific causal mutational events.

To begin to explore the contribution of environmental mutagens to cancer in humans, one must move from a microbial model to a mammalian one. Only after a connection can be firmly established between the presence of induced mutations and cancers can a mutagenic agent legitimately be called a carcinogen, or a causative agent of cancer. Scientists test mutagenic chemicals by exposing rats, mice and other laboratory mammals to them. If the application of the chemical can be correlated with an increased incidence of cancer in these mammals, the case for human carcinogenicity — cancer causation in humans — is strengthened.

But animal tests also have limitations. Cancer tests on a single mutagen can require hundreds of thousands of dollars and months, even years, of tedious studies — and even then fail to reliably predict the health consequences to humans. Only by monitoring mutations in human populations can scientists be reasonably certain of the long-term effects of chemical mutagens on the human genome. Since our society possesses deeply held values proscribing wholesale human experimentation, we are compelled to rely largely on epidemiological data — statistical patterns of cancers and other diseases in populations exposed to known mutagens — for clues. For this reason, any current catalog of documented human carcinogens will always be deceptively brief. For now, it is limited largely to X-rays and ultraviolet radiation, mustard gas, cigarette smoke and a number of naturally occurring compounds produced by plants, molds and other organisms.

But we have much yet to learn about the nature of environmental damage to DNA. We know almost nothing about the ways in which combinations of chemicals might routinely interact in synergistic ways to magnify their impact on the human genome. For example, some substances can transform a previously harmless second substance into a potent mutagen. Other substances can act in concert with a known mutagen by blocking normal DNA repair mechanisms in cells injured by the genetic toxin. Nor have we begun to comprehend the roles that mutagens might play in the still mysterious relationships between genes and between cells during the genetic symphony of human growth and development. How can we hope to design tests for mutagens that will warn us of damage to genetic processes of which we still have no knowledge?

Human Insensitivity to DNA Damage

Once one becomes aware of the expanding range of mutagenic risks our genes face almost every day, it difficult to avoid wondering why the human organism is so biologically "numb" to the molecular ravages of DNA damage. Despite the central role that genes play in our lives, evolution has left the human body strangely ill-equipped to warn us of any impending damage to our genes. Unlike so many other vital parts of our bodies — such as heart, lungs and fingertips — our genetic molecules have no sensory connections to warn our huge brains of injuries to DNA. In this sense, the human nervous system is oblivious to the hazards facing DNA molecules — whether from exposure to sunlight, ionizing radiation, certain viral infections or a lifelong barrage of chemical mutagens in the environment.

Beyond the daunting physiological problems that would have to be overcome for any recent evolutionary connection to be made between our genes and our brains, there is a sound simple reason for what at first might appear to be a serious evolutionary oversight. It is this: Evolution is incapable of anticipating the future. It can only cope with environmental challenges faced by human populations one day at a time. Our species' commitment to accelerating technological change is relatively recent. We are now introducing mutagenic agents into our environment — many of them novel, potent, concentrated and long-lived — at a pace that cannot possibly be matched by the machinery of human evolution. This does not mean that the processes of natural selection are no longer at work in modern industrial societies. Natural selection — the gradual winnowing of genes from a species' gene pool — is inevitably in motion whenever humans live, reproduce and die. But the range of evolutionary options available within the span of a few brief generations is simply incapable of equipping our species with major biological adaptations to the growing quantity of mutagenic agents we are adding to our environment.

Knowing all this, it is easy to imagine that one's genes are under some sort of massive, invisible mutagenic siege. And, to some extent, that feeling is not totally unwarranted, especially for people of those impoverished developing nations that lack even rudimentary environmental legislation to protect citizens from rising levels of genetic toxins in food and water supplies. In such situations, mutagenic substances can threaten not only somatic-cell DNA but also germ-line DNA as these substances or their by-products seep into the fetal bloodstream from chemically contaminated mothers. Genetic mutations in early embryos can do more than shorten lives and trigger somatic-cell cancers. By appearing early in the course of mitotic divisions of the growing embryo, such mutations can be transmitted to cells throughout the body — with potentially deadly effects. Furthermore, if they reach egg- and sperm-producing germ cells in the embryo, the mutation can be passed on to future generations.

Here, as in other situations we have discussed elsewhere, the price of damaging germ cells may prove to be painfully high. It has been estimated that a 20 percent increase in the number of mutations in egg and sperm cells within the U.S. population, over a period of just one generation, could have a price tag of up to 400,000 new, severe genetic defects. Presumably a similar pattern would occur in Canadian populations exposed to such a substantial increase in mutations. These inherited disorders would not surface in synchrony. An estimated 40,000 disorders might appear in the children of those suffering germ-line mutations; the rest would probably manifest themselves gradually, over a period of many generations.

How can we hope to preserve the sanctity of the human genome? Because it is so difficult to accurately quantify what many scientists acknowledge is a burgeoning threat of environmental mutation, many people seem to prefer to avoid acknowledging the problem. Others indulge in a sense of hopeless despair.

But there are at least a few more positive options. The first is to assume greater personal responsibility for understanding the nature of mutation and becoming aware of at least the major forms of potential genetic risk. The second is to try to limit personal exposure, where practical, to some of the most potent agents of mutation. The third is to encourage the development of new, more efficient tests that might be used to systematically screen commercial compounds before they enter the global marketplace. The fourth is to support governmental policies and politicians that set out to completely abolish emissions of the most hazardous, known genetic toxins — not to mention all sorts of other biologically harmful pollutants — into the environment.

Even with such common-sense strategies in place, it would be unrealistic to believe that we could possibly begin to identify every new mutagenic threat to our genes. And, considering the economic imperatives that drive most nations, it would be naive to insist that society ought to ban each newly identified human carcinogen by force of law. One need look no further than the continuing efforts to market cigarettes — despite their established role in contributing to lung cancer — to be reminded of this bitter economic truth.

There is also room for considerable optimism. We are making progress in our understanding of the mechanisms of environmental damage to DNA. In the process, we are becoming increasingly alert to the potential genetic risks of living in modern, industrial societies and increasingly adept at assessing those risks. We are, for example, discovering ways in which dietary vitamins can inactivate some mutagenic substances in our digestive tracts before they have had a chance to disturb DNA molecules.

But there are still many gaps in our knowledge of environmentally induced genetic mutations — particularly concerning the health effects of chemical mutagens and low-level ionizing radiation. As scientists continue to explore these and other issues, it is vital that they continue to communicate their discoveries to the general public. For only in this way can we hope to culturally achieve what evolution has denied us biologically — a sensitivity to the unseen sufferings of genes.

CHAPTER 11

CROSSING GENETIC BOUNDARIES: THE CURIOUS CASE OF THE CROWN GALL BACTERIUM

> **GENETHIC PRINCIPLE**
> Until we have a better understanding of the extent of genetic exchange between distantly related species in nature, we ought to consider evolutionary "boundaries" — areas of relatively limited genetic exchange — as at least provisional warning signs of potential danger zones for the casual transfer of recombinant genes between species.

The nub of the new [recombinant] technology is to move genes back and forth, not only across species lines, but across any boundaries that divide living organisms, particularly the most fundamental such boundary, that which divides prokaryotes (bacteria and blue-green algae) from eukaryotes (those cells with a distinct nucleus in higher plants and animals). The results will be essentially new organisms, self-perpetuating and hence permanent. Once created, they cannot be recalled.

> — George Wald, in "The Case Against Genetic Engineering," *The Sciences*

In a competitive natural world where genes, as well as tooth and claw, are part of a species' evolutionary arsenal for survival, the crown gall bacterium is a master of genetic guerrilla warfare. At first glance, it is a most mundane microbe, with a preference for life in the soil and a nasty habit of infecting certain species of plants with a malady called crown gall disease.

The crown of a plant is the region, located near ground level, where root turns into stem (see Figure 11.1). And a gall is a peculiar tumorous swelling that forms at the site of infection. In 1907, two scientists working at the U.S. Department of Agriculture first identified the bacteria causing this crop-damaging disease. By 1947, scientists had succeeded in artificially cultivating plant cells infected by the microbe and in establishing its cancerlike characteristics. Their work demonstrated that the injury caused by the crown gall bacterium was no ordinary wound. Rather, it was a mound of feverishly dividing, immature cells that continued to grow even after all of the infecting bacteria had been removed. Somehow, it seemed, the microbes had genetically altered cells in the plant so that they kept on reproducing.

It was another three decades before scientists began to identify the intricate genetic maneuverings of the crown gall

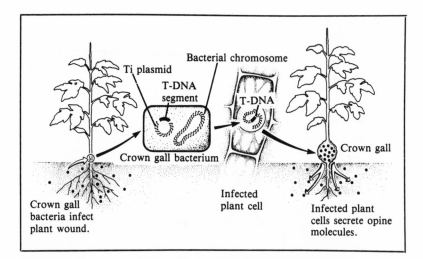

FIGURE 11.1: Life Cycle of Crown Gall Bacterium

bacterium. Even with these first revelations, few scientists could have imagined that this single-celled plant pathogen would emerge as a key to genetic engineering in crops or that it might lie at the very crux of the debate about the possible risks of indiscriminately shuffling genes between distantly related species.

The Life Cycle of the Crown Gall Bacterium

A day in the life of the crown gall bacterium is hardly a glamorous affair. In scientific circles, the parasite is known as *Agrobacterium tumefaciens*. The first, or genus, name refers to the bacterium's fondness for an agricultural habitat; the second, or species, name, to its ability to cause tumorous swellings in host plants. The minute, rod-shaped bacterium, barely visible through an ordinary light microscope, thrives in a shadowy underworld shared by a bestiary of small, soil-adapted life forms, including chlorophyll-green algae, tangles of pale fungi, marauding insects and nematode worms. In the midst of this teeming soil system the crown gall bacterium and its fellow bacteria lead a scavenger's existence, absorbing through their cell walls a smorgasbord of organic molecules that litter the soil . Over countless generations, each species has become adept at converting the atoms and energy of one or another part of this molecular debris into glistening new protoplasm. So efficient have some become that they can reproduce at an astonishing pace — splitting every half-hour or so through a process called binary fission, which is a simply choreographed version of the mitotic gene dance in your own cells.

But the crown gall bacterium's survival strategy is unique. In the darkness of the soil, it lies in wait. Its victim will be a broad-leaved plant with a fresh wound that renders it vulnerable to bacterial ambush. Normally, the stems and leaves of a healthy plant are hermetically sealed against most bacterial invasions by shielding layers of epidermal cells and tough coatings of cellulose and wax. But a plant injured by the gnashing jaws of a foraging insect, gusty winds or a disfiguring disease becomes especially vulnerable to the crown gall bacterium.

In ways still only dimly understood, the crown gall bacterium binds to a cell of the stricken host plant, then silently carries out its ingenious parasitic assault (see Figure 11.1). It begins by injecting a segment of bacterial DNA, measuring some 20,000 base pairs in length and containing a number of genes, into the injured plant. With the help of special enzymes, these foreign genes, called *T-DNA* (or transfer DNA), are stitched into a plant chromosome at an apparently random location. They then proceed to brashly issue instructions to the metabolic machinery of the hostage host cell with all the aplomb of bona fide plant genes.

But these impostor genes, it turns out, bear a selfish message. Instead of coding for products useful to the plant, they force it to make food fit for the crown gall bacterium. Called *opines*, the lightweight metabolites are never found inside a normal plant cell. But under the influence of crown gall bacterium genes, the cell now diverts some of its energy to manufacture opines that will ooze into the surrounding soil, nurturing hordes of hungry crown gall bacteria.

The crown gall bacterium is by no means the only organism in nature that so blatantly exploits the genetic apparatus of another species. Viruses, for example, routinely inject their genes into vulnerable cells in order to commandeer the metabolic machinery of their hosts. Host cells respond by churning out thousands of new infective virus particles — guaranteeing the survival of the virus. Usually such a genetic coup d'état results in the death of the cell when it swells to the bursting point with replicas of the virus. With retroviruses — viruses whose genes are written in RNA rather than the usual DNA — and *bacteriophage* — viruses that specialize in bacterial prey — the viral invasion plan is less aggressive. Instead of compelling the infected cell to commit suicide, the genes of these viruses find their way into the genetic molecules of the cell, where they may lie dormant for generations before awakening to actively code for viral progeny.

Nonetheless, in one sense the crown gall bacterium is still one of a kind. To date, the crown gall bacterium represents the only scientifically documented case in the natural world of the routine exchange of genes between a lowly bacterial cell and the more complex cells of a multicellular organism. The enormity of this feat can easily be missed. For during the course of infection, the crown gall bacterium's mobile genes are shuttled across an intercellular distance so minute it must be be measured in microns, or millionths of a meter. Yet the evolutionary distance spanned by this leap is staggeringly large. By joining a plant chromosome, the microbe's genes are in effect bridging a gap of more than three billion years — the span of time since the ancient lineage of contemporary higher plants diverged from that of modern bacteria.

Prokaryotes and Eukaryotes

To the biologist, this recently discovered display of genetic promiscuity between such distantly related species borders on heresy. The crown gall bacterium, along with all other species of bacteria and the archaic blue-green algae, is a member of what has long been considered a rather exclusive evolutionary club. Called *prokaryotes* — from the Greek word for "primitive nucleus" — these cells are distinguished by an evolutionarily old-fashioned design that preceded that of *eukaryotes*, or "true nucleus" cells (see Figure 11.2).

The most striking feature of prokaryotes is the complete absence of a membrane-bound nucleus to partition genetic molecules. Genes are arrayed in simple, unadorned loops of nucleic acid, called *chromophores*, that usually drift freely in wide-open cytoplasm, unrestricted by a nuclear sac. In the eukaryotes, which include all cells on earth besides prokaryotes and viruses, genes are arrayed in mobile, chromosomal ropes of nucleic acid and protein, which are corralled within a distinctive nuclear membrane.

The genetic molecules of eukaryotes and prokaryotes also differ in more subtle ways. Eukaryotic DNA, for example, is braided with protein. Its linear message of nucleotide base pairs is sprinkled with genetically silent segments — as well as with brief, highly repetitive sequences that, if genes could talk, we might

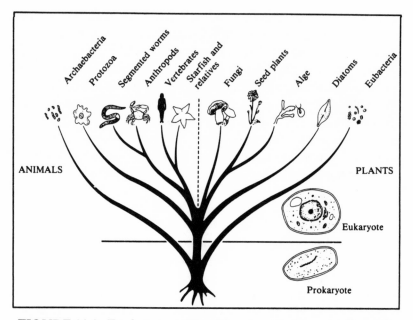

FIGURE 11.2: Evolutionary Tree Showing Prokaryotes Emerging before Eukaryotes

refer to as stutters. Its genes have distinctive promoter signals to initiate the transcription of DNA into RNA. And it possesses complex control systems, still largely mysterious to geneticists, to coordinate the diverse activities of genes. Prokaryote DNA, in contrast, is naked (without protein), straight-talking, armed with its own peculiar promoter signals and subject to far simpler control mechanisms for switching genes on or off.

Finally, the bulkier eukaryotic cell comes fully equipped with all sorts of evolutionarily newfangled gear: mitochondria power plants, chloroplast solar collectors, cilia paddles, heavy-duty ribosomal machinery and other specialized cell organelles. Together, all of these adaptations and more have transformed the eukaryote cell into an efficient, high-performance machine, but one that, over time, has proved incredibly flexible in adapting to the cooperative demands of life in a many-celled organism.

For more than 1.5 billion years after the first bacteria appeared in the earth's primeval seas, prokaryotes had a monopoly on planetary life. With the emergence of one-celled eukaryotes, that quickly changed. This development, heralding the possibility of multicellular life, may have been the single most significant evolutionary leap since the origin of life. With the separation of ancient, ancestral prokaryote and eukaryote stocks, each lineage was free to evolve independently.

The evolutionary relationships between early prokaryotes and early eukaryotes are still far from certain. But there is now a growing body of evidence suggesting that the complex eukaryotic cell may have had its beginnings in the intimate, mutually beneficial association, or symbiosis, of previously separate prokaryotic species. According to this theory of the origin of eukaryotes, chloroplasts, mitochondria and other eukaryotic organelles may be living relics of once-independent prokaryotic microbes that gradually became integrated into the cytoplasm of ancient prokaryotic host cells. Thus, some of the very anatomical innovations of eukaryotic cells that fueled the evolutionary rise of early eukaryotes may have a prokaryotic ancestry.

The Mechanism of Plasmid Exchange

The crown gall bacterium genes exported during the course of infection are located not in the microbe's main gene-bearing body, the circular chromophore of DNA, but in an auxiliary DNA hoop known as the Ti (for "tumor-inducing") plasmid. Bacterial plasmids, as you may recall, are small loops of DNA adrift in cytoplasm of a cell that are capable of shuttling genes from one bacterial cell to another in what constitutes a primitive form of "sexual" genetic exchange. The crown gall bacterium actually harbors an assortment of plasmids that come in a range of sizes, but the large Ti plasmid alone bears the genes that make *Agrobacterium tumefaciens* a plant pathogen.

While the crown gall bacterium does display some evolutionarily exceptional habits, it is hardly the only organism whose survival depends on plasmids. Biologists have tracked plasmids in an assortment of species representing both prokaryote and eukaryote lines — from bacteria to brewer's yeast and maize. Wherever plasmids occur, they tend to share a number of distinctive characteristics. First, their genes are written in the same universal genetic code as "mainstream" genes — those strung along chromophores or chromosomes and inherited through normal processes of cell division. Second, plasmid DNA often replicates independently and at a different rate than do most of the genes in a cell (although the Ti plasmid, along with a number of other plasmids, happens to replicate in synchrony with the chromosomal DNA of its bacterial host). This independent replication cycle sometimes allows plasmids to act like miniature high-speed printing presses — issuing multiple copies of their genes in rapid succession as other genes languish. Third, the movements of plasmid genes are not confined to a single cell or its direct progeny, as are mainstream genes. Rather, plasmid genes can lead a highly mobile existence as they are swapped between neighboring cells — usually members of closely related species or genera.

The effects of plasmid-mediated inheritance can be dramatic. Some pathogenic bacteria, for example, possess plasmids bristling with genes that confer resistance to specific antibiotic drugs, such as penicillin, streptomycin or other conventional antibacterial medical treatments. The ease with which these plasmids, called R (or resistance) factors, are routinely swapped between bacterial cells in nature has been made painfully evident by reports of a soaring number of hardy bacterial strains that have become genetically resistant to certain antibiotic drugs. Through our overzealous use of commercial antibiotics — both as a routine medical treatment for human infections and as an additive to livestock feed — we have inadvertently selected for the survival of pathogens that have acquired resistance genes through the natural intercellular shuttling of R plasmids.

Despite their kinship with plasmids in other species, the tumor-inducing plasmids found in the genus *Agrobacterium* appear to be unique in one respect. So far they are the only plasmids in nature known to routinely exchange genes between a prokaryotic and a eukaryotic cell. It is this rare and remarkable characteristic that makes the crown gall bacterium central to our discussion of possible "boundaries" — zones of relatively low levels of gene flow — in nature.

The circular Ti plasmid is a few hundred thousand nucleotide base pairs long — roughly 200 genes. The stretch of T-DNA that the crown gall bacterium ejects from its Ti plasmid into the plant cell measures less than one-tenth that length. A few plasmid genes are thought to somehow interact with plant genes to initiate a series of events that culminates in infection. Others, lying adjacent to the mobile T-DNA section, will remain behind. Some of these code for enzymes that prune T-DNA from the plasmid loop and send it on its way to a nearby plant cell. Others code for enzymes that will later be used to digest opine molecules released by the captive cells in the gall.

Within the T-DNA shuttle itself are genes that, once integrated into the plant genome, will direct the synthesis of the contraband opine molecules. The sudden gush of fresh opines into the soil surrounding the gall will, interestingly enough, also spur nearby Ti plasmids to replicate. These in turn, will find their way into the horde of hungry crown gall bacteria congregating near the wound — turning once harmless crown gall bacteria into new waves of infective pathogens. Other genes apparently code for the manufacture of functional plant hormones that can disrupt the plant's delicately balanced growth, sending its cells into the chaos so characteristic of cancers. Activated by the intruding bacterial genes, the renegade cells enter a state of reproductive frenzy that renders them immortal — even in the absence of bacteria.

True to the unwritten laws governing all parasites, the bond between the crown gall bacterium and the plant is a boon for the bacterium, a burden (though not necessarily a lethal one) for its host. The microbe manages to transform a living plant into a food bank, oozing opine molecules that provide a vital source of carbon and nitrogen to the bacterium. Genetically shackled by the Lilliputian germ, the plant continues to convert sunlight into opines, nurse its festering wounds and struggle to survive. But the crown gall bacterium has also introduced a novel twist to its parasitic role. It alone among bacteria possesses the enzyme-coding genes needed to devour the opines. So evolutionarily fine-tuned is its assault that it has managed to outmaneuver any competitors that might try to share in its spoils. Not only has the microbe genetically conscripted its meal, it has made certain that there will be no other freeloaders at its solitary feast.

Implications for Genetic Engineering

The same qualities that allow the crown gall bacterium to carry out a genetic coup d'état in plant cells make it a useful tool for genetic engineering applications. The species is already a proven professional when it comes to splicing genes. The crown gall bacterium comes fully programmed not only to deliver genes into a target plant cell but also to integrate them into the clockwork harmony of resident plant genes.

Furthermore, the transspecies transport of genes carried out by the crown gall bacterium is also part of a larger evolutionary system that is not always easy to mimic in a laboratory. Through a painstakingly slow interaction with their eukaryotic hosts, crown gall bacteria have undergone what might be called a cooperative natural selection with their plant hosts. Through their continuous evolutionary relationship, host and parasite have forged an intimate and a long-standing bond. At times, it takes the form of what might be called a genetic "dialogue" — an adaptation, mediated by a single gene, in one species is "answered" by a corresponding single-gene innovation in the other. In this way, genetic exchanges across the hypothetical prokaryote-eukaryote boundary occur in minute increments — one gene at a time — without disturbing the delicate balance of either species.

The intricate metabolic processes of eukaryotic cells can be easily disrupted by the haphazard insertion of foreign DNA sequences into their highly organized genetic molecules. For this reason, the seasoned crown gall bacteria, with their extraordinary ability to stitch plasmid-borne genes into the genetic apparatus of eukaryotic organisms, are almost ideally suited to deliver recombinant DNA sequences into a eukaryotic genome. All one would have to do is attach newly cloned genes from a third species — one harboring genes coding for a desired trait — to the T-DNA shuttle in its Ti plasmid (see Figure 11.3). Laden with this recombinant DNA, the T-DNA fragment would depart from the plasmid like a locomotive from its station. From there it would travel along an enzyme-built railway into the injured plant cell — in a laboratory-grown plant, for instance. Finally, it would release its load of imported genes to find a resting place amid the coils of eukaryotic DNA. If any of the smuggled genes coded for detectable bacterial traits such as drug resistance, so much the better. They could serve as markers to confirm that a given stretch of DNA not only had made it into the host but had worked its way through the maze of metabolic machinery leading to protein.

The shopping list for potentially useful genes to shuttle into plants seems positively endless. Why not, some have suggested, infect a corn plant with crown gall bacteria bearing genes that code for the handful of amino acids needed to transform kernels into nutritionally complete sources of protein like meat? Why not shower wheat fields with crown gall bacteria equipped with cloned nitrogen-fixing (*nif*) genes from a bacterial relative of the crown gall bacterium called *Rhizobium* — an ancient microbial ally of farmers that has evolved a symbiotic existence in the roots of beans, peas, lentils and other leguminous crops? This battery of high-powered genes enables *Rhizobium* to pluck airborne nitrogen gas from the atmosphere and chemically convert it into heavier soil-bound molecules that, readily absorbed by plants, represent a valuable fertilizer. By circumventing the bacterial intermediary and inserting these potent genes directly into wheat cells, one might be able to transform wheat plants into living wheat-fertilizer factories. All any scheme seems to require is that one find a way to identify, isolate and clone the desired gene — then one simply places it on board a crown gall bacterium and sends it into the agricultural plant of one's choice.

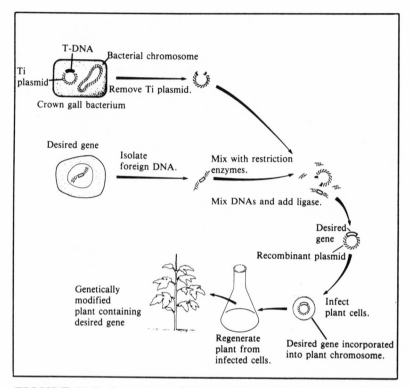

FIGURE 11.3: Insertion of Recombinant Genes into Tobacco Plant Using Crown Gall Bacterium Vector

Technical Problems

But those tasks have proved more challenging than many scientists had at first supposed. It soon became clear that many of the traits most valued in agricultural crops arose not from one or two genes but from the interaction of tens or even hundreds of genes. For instance, there may be more than eight precious *nif* genes in a *Rhizobium* bacterium. And they refuse to fix nitrogen except in a harmonious chorus. Another species of nitrogen-fixing bacteria depends on 17 separate *nif* genes, operating under an elaborate control system, to carry out the task.

Unless oxygen levels in the new host cell were finely tuned, a critical enzyme coded by one of them might be rendered useless. If most important traits in crops were polygenic, how could one ever hope to track down every component gene (especially ones that left no easily detectable protein trace), let alone remove and reassemble them in a foreign cell, without disrupting their delicate dances?

Furthermore, before it could serve as a suitable vehicle for transporting recombinant genes, the crown gall bacterium's Ti plasmid would require some laborious customizing. The nasty T-DNA genes that plunge infected plant cells into a pathological cancerous state would have to be pruned away. Otherwise bacteria might leave fields of tumor-infested recombinant crops in their wake. Once they have been removed, altered plant cells could be carefully cultivated in laboratory glassware containing a broth of nutrients and hormones designed to nurture them into healthy, full-grown recombinant plants. With a little luck, some of them might even transmit to their offspring copies of the new genes by way of the long-lived germ plasm of their seeds.

In addition, short signal segments of DNA crucial to the control of gene transcription and translation must flank the main text of the desired gene carried by the Ti plasmid. These serve as genetic punctuation marks, ensuring the recognition and expression of the foreign gene by the eukaryotic metabolic machinery of the plant cell. One DNA sequence, called a eukaryotic promoter, controls transcription of the desired gene into plant messenger RNA. The other DNA sequences consist of start and stop codons, which initiate and terminate translation of the messenger RNA by the plant cell's ribosomal protein factories.

Before transporting a passenger DNA sequence into a plant, one removes the cluster of crown gall bacterium genes occupying the mobile region of the Ti plasmid, leaving critical control sequences attached to the plasmid loop. Then the desired gene is extracted from a third species and is simply inserted next to the transcription control sequence and between a stop signal and a start signal. Without this preparatory step, the foreign DNA sequence would remain metabolically mute on reaching the host plant cell. With it, the gene is granted passage from the pro-karyotic world of the crown gall bacterium to the eukaryotic kingdom of plants.

Variations on this basic strategy have met with a number of impressive, but qualified, successes. In 1981, genetically altered bacteria were successfully grown in mature plants. A year later, genes from species as diverse as bacteria, mammals and plants had been inserted into other plant hosts using crown gall bacteria — although only those from plants were expressed. Since then, scientists have learned how to routinely enclose desired "passenger" DNA segments within a parentheses of appropriate promoters and other signal DNA sequences to facilitate their expression in plant cells. In some cases, the desired recombinant genes are not only expressed in their new host but also remain stably integrated within plant chromosomes throughout the gruel-ing ordeal of meiosis. Thus, they are transmitted into seeds and can find expression in the subsequent generation. Whether it will prove possible to ensure the long-term stability of recombinant genes in the germ-line cells of domestic crop species remains, for now, uncertain. But, based on studies revealing that lingering genetic scars of natural *Agrobacterium* infections seemed to have had no untoward effects in a number of plant species, many researchers remain optimistic.

Until recently, another fundamental obstacle to the genetic engineering of crops also remained. Except for the eccentric crown gall bacteria, very few species are available that might be har-nessed as vectors to transport desired genes into domestic plants. Alternative vectors were desperately needed because the crown gall bacterium's capabilities appeared to be extraordinarily limited. It seemed to be evolutionarily adept at inserting genes only into the genomes of host species categorized as dicots — flowering plants characterized by broad leaves and seeds with distinctively paired cotyledons, or seed leaves. Thus, the cells of fruit trees, grapes and tomatoes for instance, are potential targets for the crown gall bacterium's mobile Ti plasmid. But most of the world's economically important crops — including cereals and grasses — happen to be monocots, plants with narrow, parallel-veined leaves and only a single cotyledon, or seed leaf. As a result, unless the natural host range of the crown gall bacterium could somehow be artificially expanded, these prime targets for future genetic enhancement appeared to be beyond the genetic engineer's grasp. It now appears that a number of the world's most economically important plants, including rice and corn, can, in fact, be transformed by *Agrobacteria*. Thus, the potential value of the crown gall bacterium in the genetic engineering of agricultural crops is likely to greatly exceed earlier pessimistic estimates.

There are a few intriguing candidates for alternative vectors, among them, a viral pathogen, known as the cauliflower mosaic virus, and a bacterial pathogen, a relative of *Agrobacterium tumefaciens*'s that instead of triggering crown galls causes a proliferation of opine-laden roots. But, for now, neither seems as promising as the crown gall bacterium.

Questions about Future Applications

If scientists ever do succeed in harnessing the crown gall bacterium and other species reliable as gene-bearing beasts of burden in the genetic manipulation of domestic crops, they could usher in a new and revolutionary era in agriculture. In principle, they would be placing the genetic resources of the entire biosphere at the service of farmers. As a result, the potential benefits for humans from the genetic manipulation of crops could be beyond calculation.

As geneticists unravel the hereditary mechanisms underlying critical metabolic processes in plants, they could seek to improve these mechanisms. They might try to tinker with genes controlling photosynthesis, for instance, in order to create new "super-crops" capable of converting sunlight into plant tissue with increased efficiency. They might try to genetically tailor crops to tolerate saline or acid soils that currently lie uncultivated around the world. Or they might attempt to manipulate cellular biochemical reactions in ways that matched the nutritional content of food harvests to human nutritional needs.

But, as in most other arenas of genetic technology, scientific discoveries are also bound to present many difficult moral questions. Some of the most fundamental questions will concern which applications to pursue. When the future directions of genetic engineering in agriculture are shaped by private-sector as well as by public-sector interests, how will we ensure that the results are humane and fairly distributed? Will the goals of projects aimed at genetically altering food crops, for example, be to satisfy the pressing nutritional and economic needs of the vast majority of human beings who share our planet? Or will they instead be dictated by the often unfeeling forces of the marketplace, as innovative young biotechnology firms, staffed by some of the best minds in science, compete to satisfy the short-term interests of their shareholders?

Today, for example, one of the top priorities among biotechnology companies engaged in agricultural gene-splicing projects is to genetically improve economically important plant species by inserting genes coding for enzymes that make crops resistant to certain commercial herbicides. If this feat were accomplished, pesticides would kill weeds and other harmful competitors but leave genetically resistant recombinant plants unharmed. The economic rewards of growing such genetically modified crops — reflected in larger harvests — might be difficult to ignore. But would the cultivation of genetically modified, pesticide-resistant species also conceal the disturbing rise in levels of toxic farm chemicals in our agricultural soils and surrounding ecosystems? Is it wise to entrust the development of such genetically engineered crops to the same corporations that profit from the very pesticides that pollution-resistant plants would be designed to endure? And by harnessing genetic technologies in such a manner would we be committing ourselves to a never-ending cycle of searches for technological fixes to solve failures in our earlier technological fixes?

Other important questions arise regarding the possible long-term consequences of releasing commercial recombinant plants into the wild. For example, by splicing genes from distantly related organisms, could we inadvertently create life forms that by escaping an ecosystem's natural controls, might create ecological havoc? Could foreign genes inserted into domestic species occasionally interact with the host genome in unexpected ways to produce cellular substances toxic to those who consumed them? Could overzealous efforts to customize pathogens such as the crown gall bacterium by extending their host range inadvertently lead to outbreaks of disease in previously unaffected plant species? By too hastily splicing genes coding for herbicide resistance or another similar trait into our crops, could we accidentally leak those DNA sequences into the genomes of adjacent fields of related noneconomic species that could then genetically sidestep chemical control? The risk of introducing such genes into surrounding weed populations, for example, is compounded by the fact that recombinant DNA sequences could be broadcast in clouds of airborne pollen. Thus, sowing certain genetically modified crop species could literally mean throwing foreign genes to the wind.

For now, at least, our scientific understanding of the possible long-term effects of releasing novel recombinant organisms into ecosystems around the world is simply too crude to permit us to reliably answer these and other unsettling questions. As a result, any discussion of the potential risks of agricultural genetic engineering is necessarily speculative — as are most confident assurances that no risks exist at all. It is also fair to argue, of course, that we will never know enough about the ecological impacts of recombinant creatures to reduce the risk of their release to zero. It will be up to society as a whole to determine precisely what levels of environmental risk we are willing to tolerate to reach our short-term economic goals.

The Nature of Genetic Boundaries
Meanwhile, what are we to make of *Agrobacterium tumefaciens?* How important is its seemingly solitary breach of what was once

considered at least a conceptual barrier to routine genetic exchange between prokaryotes and eukaryotes? Is it an isolated case — a mere trickle of genes? Or is it just the first scientifically documented sighting of what will prove to be a much larger traffic in genes between distantly related species that for now is beyond our ability to detect? And if it turns out to be the latter, are we even justified in postulating genetic "boundaries" in the natural world? For if the crown gall bacterium, retroviruses and other, as yet unknown agents routinely carry out the exchange of genetic information between distantly related species, perhaps nature really has no evolutionary barriers to breach.

Far more hinges on these questions than one might suspect. The crown gall bacterium is seen by some scientists as living proof that nature is tolerant of genetic exchange between even the most distantly related species. Crown gall bacteria, so the argument goes, have been testing the prokaryote-eukaryote boundary for millennia, with no harmful results. In this sense, the crown gall bacterium stands as the ultimate genetic entrepreneur — an enterprising prokaryote pathogen that reveals the possible evolutionary rewards of exploiting eukaryotic genomes.

How many cases like crown gall are there in the natural world that have simply eluded human detection? No one knows for sure. Compared with the mainstream currents of genes that flow vertically through time from one generation to the next, any horizontal trickles of genetic information between species may simply be too minute to detect. But many researchers are convinced that horizontal gene transfer between taxonomically distant species, however elusive for the moment, will eventually turn out to be a far more common genetic process than modern evolutionary theory currently suggests.

Even now, scientists do know, for instance, that bacteria of one species routinely exchange genes with bacteria of other species. To accomplish this, they have come up with several effective techniques. Some bacteria can send genes from one cell to the next by a transient bridge of cytoplasm. This quasisexual process of genetic exchange is known as *conjugation*. Other bacteria conceal genes, like soldiers in a Trojan horse, within the genomes of obliging viruses that prey on many bacterial species.

Within the confines of the laboratory, some bacteria have been coaxed to directly absorb and make use of naked DNA from their surroundings. This classic observation has led some scientists to suggest that bacteria exposed to eukaryotic DNA in nature — for example, bacteria of decay on the carcass of a dead eukaryote animal — might secretly have been incorporating eukaryote genes for centuries. If so, these scientists continue, every conceivable combination of prokaryote and eukaryote genes must have been tested — and proved harmless — over eons of natural selection.

This may be a comforting thought to some gene splicers, but it remains largely speculation. While in some instances bacteria absorb DNA directly from their surroundings and then go on to express and transmit these foreign genes, there is little evidence yet that the cannibalization of raw DNA is a routine avenue of genetic exchange in nature or that it enhances bacterial survival.

When natural gene exchanges do take place among bacteria, they are usually restricted to members of closely related species or genera. And boundaries between bacterial populations are often rigorously patrolled at the molecular level. Ironically, one of the most powerful tools in recombinant DNA technology — restriction enzymes or bacterial enzymes capable of cutting DNA molecules at specific base sequences — has been derived from a biological mechanism by which bacterial cells defend their genomes against intrusions of foreign DNA.

Within a living bacterial cell, teams of resident restriction enzymes form a sophisticated natural defense system that can distinguish foreign from homegrown DNA. With their finely tuned chemical specificity, these enzymes collectively patrol the bacterial cell cytoplasm for signs of alien DNA molecules. If intruders are present, the restriction enzymes swoop down and bond to recognizable sequences, simultaneously slicing the genetic molecules into nucleic acid ribbons. Because the bacterium's own DNA is protected from restriction-enzyme attack by an ingenious chemical armor, it is left completely unscathed. The effect of such enzymatic search-and-destroy missions is to help maintain the genetic separation of certain prokaryote populations in much the same way that differences in sexual behavior, anatomy of sexual organs, timing of reproductive season or other species-specific traits serve to limit genetic exchange between unrelated species of sexually reproducing animals.

Furthermore, the textbook notion of a species as a single inter-breeding population of organisms has meaning only in the context of sexually reproducing species. The definition becomes strained in the realm of bacteria, which reproduce predominantly by asexual means. Lacking the true forms of sexual reproduction mechanisms found in higher organisms and bristling with promiscuous gene-bearing plasmids, prokaryotic species tend to form somewhat more fuzzy "species" divisions than their eukaryotic counterparts. As a result, a trespassing of boundaries within one group does not necessarily correspond to the same event in the other. And any reassuring references by scientists to the precedent of natural gene transfer between species of bacteria ought to be judged in that light.

The Crown Gall Bacterium as Nature's Exception
If the crown gall bacterium retains its singular status as the only prokaryotic species to exchange genes with eukaryotic genomes, it could, paradoxically, become a symbol for an entirely opposite view. From a broader evolutionary perspective, its unique hereditary habits could be seen as compelling evidence not for nature's laissez-faire acceptance of genetic free trade, but for the astonishing rarity with which prokaryotes and eukaryotes routinely shuffle their genomes.

In fact, the crown gall bacterium could be seen as the exception that "proved" the widely accepted "rule" among evolutionary biologists that nature tends to limit the flow of genes between distantly related species. This rule is rooted in the notion that the main impetus of natural selection is the gradual establishment of biological barriers, which is the means by which the ancient machinery of evolution generates new species. Over time, heritable traits that enhance the reproductive success of individual organisms in a species are favored by a particular environment and accumulate in a pool of genes. But, at the same time, a variety of processes are put into place to insure that hard-won survivor genes are not quickly diluted by the genes of neighboring populations.

Biologists refer to these obstacles as reproductive isolating mechanisms. Their effect is to limit gene flow by foiling illicit matings between species — especially distant relatives. Any trick that restricts reproduction to bona fide members of a species' gene pool can serve this function. The intimate clockwork timing of pollination in flowering plants or the aquatic courtship dance of breeding pairs of waterfowl confine precious germ plasm to members of the same species. Likewise, the precise anatomical fit, like lock and key, between the chamber of an orchid's flower and the bristly body of an insect that has coevolved for centuries to efficiently pollinate it serves to limit genetic exchange with evolutionary outsiders.

Sometimes a barrier against genetic exchange between species is nothing more than a mountain, a river, a temperature gradient or an isle-encircling sea. All it must do is block interspecies gene flows. When all such measures fail to prevent the critical fusion of egg and sperm, other less visible mechanisms can come into play. A rare hybrid offspring of two parental species, for example, may reveal an internal chaos of nonhomologous chromosomes in premature death, infertility (as with the mule) or the progressive weakness of future generations.

Still, natural barriers to DNA exchange are not necessarily absolute. Reproductive isolating mechanisms sometimes cease to act as obstacles to gene transfer when two species are brought together under artificial conditions. And even under natural conditions species may be separated by no more than a statistical gene barrier — the low mathematical probability of interspecies genetic exchange.

The potential widespread release of novel recombinant organisms poses a possible threat to the integrity of this vast evolutionary pattern, which biologists refer to as the "continuity of divergence" in evolution. The phrase is another way of saying that the principle thrust of evolution has been to gradually establish a diversity of separate gene pools and to test them against an ever-changing environment — without ever allowing them to coalesce again. It is this inexorable push to diversify that results in species, genera, families and other taxonomic classifications that contribute to the evolutionary tree of life.

In its wake, this process has left a legacy of distinct lineages of living species that while sometimes blurred at their edges relentlessly continue to diverge. That is, the limbs of the 3.3-billion-year-old evolutionary tree do not branch outward only to fuse together — in the way evolving human languages sometimes rejoin to create a new dialect. Once species arise from an ancestral stock, they follow independent paths — never fusing to form a new species. Nor is evolutionary history ever repeated. Our image of evolutionary history as a tree underscores this easily forgotten truth. If natural selection rewarded wholesale gene exchange between distantly related organisms, we would be compelled to find another, more suitable model of evolution. We might, for example, choose to depict the history of life on earth not as some ancient oak but rather as turbulent ocean waters or a boiling kettle of tea.

Lessons from the Crown Gall Bacterium
If the crown gall bacterium offers us few final answers, it does at least provide us with an intriguing glimpse of a prokaryotic lifestyle that successfully straddles the very prokaryote-eukaryote boundary that new genetic technologies bid us explore. But does it indicate whether we ought to casually cross that fundamental — if not absolute — evolutionary border by releasing endlessly imaginative genetic novelties into our surroundings?

The very idea of searching for ethical meaning in the natural history of a humble microbe is, of course, heretical to science. For the scientific worldview tends to be dominated by the conviction that questions of morals and questions of science occupy two strictly separate philosophical provinces. Many scientists perceive the natural world as morally neutral and utterly devoid of messages or clues that might inform questions of human ethics. For this reason, neither the crown gall bacterium nor any other piece of nature's handiwork can possibly serve as a moral beacon to warn us away from possible biohazards or underscore evolutionarily ordained "oughts."

There are, of course, many cultural traditions around the globe — from the myths honored by the Haida Indians of the Pacific Northwest to the teachings of Tibetan Buddhists — that would challenge the scientific notion that nature is morally neutral. But in the context of scientific objectivity, perhaps the only way that the natural world can offer us moral guidance is through metaphor — imaginative, insightful leaps of analogy. And that may be precisely the role that *Agrobacterium tumefaciens* can play.

For in the crown gall bacterium, nature presents us with an evolutionary paradox. The microbe represents a strikingly successful evolutionary experiment. Yet, at the same time, its long-distance gene-foolery borders on the irreverent — flaunting as it does the very boundaries between major taxons that other assorted evolutionary metaphors have taught us to respect. It is the apparent *rarity* of such prokaryote-eukaryote gene transfer that may, in the end, transform this microbe into a metaphorical warning. The message may be simply that despite the strange case the crown gall bacterium the predominant pattern in evolution remains that of a "continuity of divergence" between prokaryotic and eukaryotic gene pools. And if this message is true, perhaps before disrupting this ancient evolutionary architecture on a large scale, we ought to explore nature's biological "motives" for its vigilant and widespread maintenance of genetic borders between distantly related species.

Perhaps unfairly, we have chosen to postpone any mention of another recent, and still somewhat enigmatic, bit of scientific evidence concerning *Agrobacterium tumefaciens* that adds another layer of ambiguity to this microbe's uncertain moral metaphor. We had two motives for the delay. First, unlike most of the scientific data concerning the crown gall bacterium already discussed in this chapter, the meaning of this information remains highly speculative. Second, by waiting until late in our story to reveal it, we may be able to cast a completely different light on the possible significance of *Agrobacterium tumefaciens* and thereby discover a broader, more philosophical genethic lesson.

The news is this: Closer scrutiny of T-DNA sequences carried by the microbe's Ti plasmid has recently revealed that these are strikingly "eukaryotic" in their layout and punctuation. That is, they are chemically more complementary to eukaryotic plant DNA than to prokaryotic bacterial DNA. And their promoters are of eukaryotic design. What this ultimately means is uncertain. But some geneticists believe that T-DNA may not, in fact, represent heretical prokaryote DNA at all. Instead, it may be an ancient shard of eukaryotic DNA, removed long ago from from a plant genome, that finds ready expression in multicellular hosts today simply by riding a plasmid back home.

There are alternative speculations too. The eukaryotic sequences in T-DNA could have arisen by statistical chance. These sequences may have been inherited from a shared ancestor in the distant past. They might even have appeared first in bacteria, then later gone on a long sojourn in plants. But if the first scenario proves true, the genetic cargo carried into plants by the crown gall bacterium may not be subversive prokaryotic genes at all. It could be nothing more than migrant eukaryotic genes, out for an extended prokaryote ride — contributing, perhaps, to the genetic diversity of plant gene pools and establishing a counter-current eukaryote-prokaryote genetic exchange in *Agrobacterium tumefaciens.*

If this is so, we may have to find a moral metaphor that better suits this maverick microbe. And to be faithful to the facts, we may have to create one with a double image. The first: that the crown gall bacterium may be, after all, living confirmation that nature does not treat lightly the presence of prokaryote genes in eukaryotic genomes. The second, more sweeping and sound in any event: that in the scintillating world of genes, we would be wise to anticipate that things are not always as simple as they at first seem.

CHAPTER 12

MAIZE: IN PRAISE OF GENETIC DIVERSITY

> **GENETHIC PRINCIPLE**
> Genetic diversity, in both human and nonhuman species, is a precious planetary resource, and it is in our best interests to monitor and preserve that diversity.

> *On the fourth day, as the sun turns towards the West, the wizards . . . grant them passage to the flatlands, where the maize in all its forms awaits them, in the flesh of their children who are made of maize, in the bones of their dead ones, who are skeletons of maize, powder of maize. . . .*
> — Miguel Angel Asturias, in *Men of Maize*

It would be hard to find a more fitting image of the stunning successes of twentieth-century agricultural genetics than a field of Iowa corn, stretching to the horizon in a rustling sea of green, as it ripens under an August sun. Yet that same scene could serve equally well as an image of one of agriculture's most perplexing problems — the steady loss of genetic diversity within many species of domestic crops.

If you could peer into the nucleus of a cell in every corn plant in such a field, you could see actual evidence of this disturbing trend. Nestled among thousands of genes strung beadlike along the 10 pairs of maize chromosomes, you would eventually come across a number of genes that were identical in almost every corn plant in sight. Most of these shared DNA sequences, painstakingly bred into this variety of hybrid corn by plant geneticists, contribute to structural or physiological characteristics that add to the economic value of the crop — uniform height and kernel size for ease of harvesting, optimal tolerance to temperature and soil conditions, resistance to particular diseases.

These and other inherited traits have contributed spectacularly to the increase in yields brought about by the mid-twentieth-century Green Revolution in agriculture. And this pattern of genetic uniformity within the maize gene pool extends far beyond this Iowa cornfield, encompassing large areas of North America, Asia, Africa, Europe and other regions of the world where agricultural techniques based on *monoculture* — the widespread cultivation of genetically similar plants — has established vast continental carpets of modern maize.

What would not be visible in this bountiful scene — unless, by chance, this particular crop happened to be engaged in a life-and-death struggle with some corn pest that had managed to penetrate its rigid wall of genetic defenses — is the possible long-term price of such homogeneous genetic landscapes of *Zea mays*, or maize, as it is commonly called throughout most of the world.

The Nature of Maize
Maize is an exceptional plant. So adept is it at capturing sunlight and converting it into broad, parallel-veined leaves, thick stalks and swollen, seed-bearing ears cobbled with neat rows of kernels that it now ranks as one of the most efficient agricultural crops grown in temperate climates. A single corn seed, sown one centimeter deep in rich prairie soil, nursed by a regimen of chemical fertilizers and pesticides and groomed by a fleet of farm machinery, weighs no more than half a gram. Yet in the span of a few months, it can grow into a towering three-meter-high plant, laden with ears that may each hold several hundred new seeds. That prodigious growth rate has catapulted a species that less than 500 years ago was unknown outside American Indian

societies in the New World into the third most important agricultural crop (after wheat and rice) in the world and the single most important domestic plant species in many nations, among them the United States.

Today there are six main types of maize, encompassing a large number of varieties or races — all bona fide members of *Zea mays* (see Figure 12.1). Each type is sufficiently different from the others to supply its own distinctive set of products.

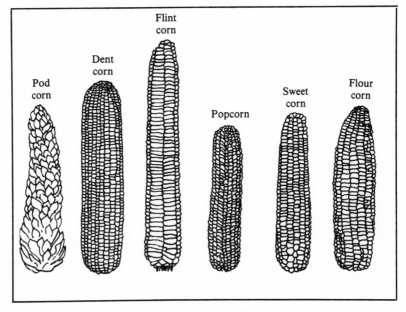

FIGURE 12.1: Cobs of Six Types of Maize

1. Dent corn can be distinguished by kernels that have a small surface pit, or dent, on their outer surfaces. It makes up the majority of the world's annual corn crop — fed mostly to livestock, which can convert it to the higher-grade proteins of meat.

2. Sweet corn, characterized by the sugary juices in its ripe kernels, is familiar to most North Americans as the table vegetable we call corn on the cob.

3. Popcorn, as we all know, possesses kernels — hard shelled, angular and moist — that explode inside-out at high temperatures, turning into the crunchy puffed snack we associate with movie theaters.
4. Flour corn has a softer, more powdery store of starch in its kernels. It is the principal ingredient of tortillas and other fire-parched unleavened corn breads that have over the millennia fed American Indian peoples ranging from the Mound Builders of the Mississippi valley to the great Maya, Aztec and Inca civilizations of Central and South America.
5. Flint corn typically has tough, oblong kernels that thrive in cool northern soils.
6. Pod corn, the most primitive of the six types and a throwback to ancient forms of cultivated maize, has virtually disappeared. It can be recognized by its distinctive, evolutionarily old-fashioned habit of wrapping each kernel in a papery protective leaf, or bract.

Bred to match the mainstream tastes, milling machinery and marketing strategies of wealthy industrial economies, these six families of maize have by now supplied us with everything from breakfast cereals and salad oils to vats of amber, well-aged whiskey.

The Origins of Maize

The origins of maize are shrouded in mystery. But evidence suggests that some species of wild plants had already been domesticated by ancient American Indian farmers nearly 8,000 years ago. Remnants of domestic corncobs, some dating at least as far back as far as 4,000 years ago, have been unearthed in archaeological sites in New Mexico, the Tehuacán Valley south of Mexico City, and highland Peru, indicating that domestication of this New World grain was well underway by that time.

Judging by the relatively large number of genetically diverse local varieties of so-called Indian corn discovered in the region, Central America is a likely center of early maize cultivation. A second is located along the eastern slope of the Andes mountains in the vicinity of present-day Peru. Whether these represent sites of independent maize domestication or simply the secondary surfacing of a single primal seed stock is not certain. In any case, seed from both Latin American centers of maize diversity probably radiated out, like ripples on the surface of a stone-shattered pond, along overland trade routes, eventually taking root in habitats ranging from the cool shores of Canada's Gulf of Saint Lawrence to the parched deserts of the American Southwest and South America.

Despite decades of sleuthing, botanists have not yet managed to fully document the evolutionary history of maize. Nor have they identified with certainty the species that were first modified by early agriculturalists to become *Zea mays*. Nonetheless, this search remains an essential task in the long-range conservation of genetic variability in maize.

If remnant populations of the wild progenitors of maize still survive, they would probably be found near the Central or South American focus of maize domestication. But some scientists suspect they may already have suffered extinction — either disappearing thousands of years ago from natural causes or vanishing more recently, like some primitive relatives of maize, through human destruction of their habitat.

Two species of wild grasses have been found in Central America that are so closely related to maize they can successfully interbreed with it. The cells of one, called teosinte after its original Aztec Indian name, resemble maize in having 10 pairs of chromosomes. Teosinte plants often grow as wild weeds adjacent to Central American cornfields and, while strikingly different in form, show some tantalizing structural similarities to maize. They have, for instance, small cornlike ears, each bearing about a half-dozen kernels. But teosinte otherwise remains genetically far removed from its docile relative.

The second grass, *Tripsacum*, has 18 pairs of chromosomes and seems even more distantly related to maize. Current speculation, though far from unanimous, suggests that a teosintelike form may, in fact, have been ancestral to domestic maize, which, in turn, later cross-bred with *Tripsacum* and other wild species that lingered near the edge of tilled fields.

But does it really matter whether we ever uncover a wild ancestor to maize or not? And what would be the practical advantages of finding and preserving this ancestor?

First, such a discovery would grant plant breeders access to a reservoir of genes closely related to modern maize. Among them, one suspects, would be genes that, though gradually lost from the gene pool of maize during the centuries of its domestication, could code for characteristics that might help maize adapt to new agricultural settings — without disrupting the finely tuned interactions of existing genes in a maize genome.

In recent years a discovery was made that underscores this possibility. On a hillside in Mexico, scientists chanced upon what turned out to be a scientifically unknown variety of wild corn — now known as *Zea diploperennis*. Unlike its domestic cousin, it is a perennial plant and therefore does not depend on a fresh sowing of seed each season. It also displays genetic resistance to a number of viruses and seems to thrive in wet soils. The possible benefits of somehow transferring such economically desirable traits into the gene pool of commercial maize varieties are obvious. But the genes of *Zea diploperennis* hang in precarious balance — only a few thousand of these wild corn plants are now known to exist.

Second, locating a wild form of maize with a relatively intact gene pool, untouched by human hands, might also grant us a clearer vision of the history of maize. It might help botanists stage crude reenactments of the unrecorded evolutionary events by which humans — consciously or unconsciously — systematically selected the hereditary traits that resulted in its domestication. Additional evolutionary clues could be gleaned by comparing the DNA sequences of maize and its ancestors.

The Domestication of Maize
Today speculative scenarios of the domestication of maize seem

almost mythical in their simplicity. Presumably the process begins with the accidental discovery by one or more groups of early hunting-and-gathering peoples that seed from a wild edible plant thrives in the nitrogen-rich garbage heaps near their settlements. By harvesting this conveniently situated wild crop, sowing seed from its most vigorous plants next season and perhaps plucking a few weeds from the primal garden, artificial selection of maize can be said to have begun.

At first, the process of artificial selection need not have been deliberate and methodical. It might well have been more unconscious and haphazard, including natural preservation of the most valued portion of the crop and the destruction of the less valued portion. Nor was it likely to have occurred in an agricultural vacuum. While maize was one of the pillars of the Aztec, Maya and Inca civilizations, New World peoples pioneered the domestication of a long list of other crops — among them, the ancestors of modern tomatoes, beans, squash and potatoes.

Domestication linked humans to maize. By artificially selecting for a constellation of desirable traits in their first crops — among them, rapid growth, a brief synchrony of sexual activity, big multikerneled ears and readily harvested cob-bound seeds — early farmers forced the wild ancestors of maize to adapt quickly to the dramatically different environmental conditions of the agricultural field. Natural selection suddenly favored those genotypes that were best adapted to survival in a modified life cycle characterized by hand-sown seed, artificially tilled soil and periodic human harvests of seed.

By domesticating maize, early farmers, in a sense, enslaved it. Centuries of artificial selection have effectively paralyzed the seeds of maize. Since maize kernels have been bred from the beginning to remain firmly embedded in the cob to facilitate their harvest, they have all but lost their ancient ability to disperse individually. The species can survive only if human hands break the kernels free and transport them safely to new soil.

In an agricultural setting, human labor enhanced the survival of maize plants by improving growing conditions, limiting natural genetic exchange with other related species and carrying out a portion of the maize reproductive process by hand. As a result, the bond between domesticator and domesticated developed into a mutually beneficial evolutionary marriage — for better or worse — that in biology is called symbiosis.

The Bond between Humans and Maize

So intimate was the partnership between humans and maize in some ancient American Indian societies that maize took on a distinctly religious aura. Special ceremonies accompanied the planting, cultivation, harvesting and milling of maize. Images of the plant adorned some of the finest pottery created by Indian artisans (see Figure 12.2).

During the Inca Empire's Classic epoch, beginning about a thousand years ago, the Inca measured the passage of time with maize. Their rulers developed a graphic agricultural calendar based on the life cycle of maize — an open acknowledgment of the pivotal role this plant species played in their civilization. Among the mythical creation stories of the ancient Maya of Central America was the belief that maize had existed long before humankind. The very flesh of human beings, they thought, had been fashioned from the blending of divine cornmeal and serpents' blood.

Is it any wonder, then, that aboriginal farmers treated maize with a reverence and respect that is completely foreign to our own scientific worldview? Sometimes, in reading a modern scientific monograph on maize, one encounters a brief reference to the enormous affection that some ancient American Indian societies apparently lavished on *Zea mays*. The notion of humans openly proclaiming love for corn plants is usually relegated to romantic anecdote — one lacking in any evolutionary significance whatsoever. Yet this ancient bond between human beings and maize — so heretical to the sensibilities of mainstream modern science — raises some intriguing questions.

There is little to suggest, for example, that the wild profusion of red and purple kernel colors that distinguish some races of

FIGURE 12.2: Mayan Maize-Bearing God with a Headdress of Cobs

Indian corn served any pragmatic purpose. Perhaps this carnival of colors represents nothing more than an evolutionary accident. But it might reflect something more. Could early maize growers have selected such traits in a utilitarian crop for strictly aesthetic reasons? Could these colors have been regarded as a sort of sacramental ornamentation symbolizing aboriginal devotion to the spiritual values embodied by their beloved crop?

Today plant geneticists seem to suggest that they have captured the essence of the striking hues of Indian corn kernels within sterile, mathematical equations describing the statistical interplay of four separate genes on maize chromosomes. But could such flaming colors also have arisen as a silent expression of interspecies gratitude? Is it possible that science simply has a built-in blind spot to the extraordinary love that members of certain nonscientific societies have lavished on the life forms around which their own lives revolve?

The Making of Modern Maize

All six of the major lines of modern maize already flourished in the New World on November 5, 1492, when two Spaniards sent out by Christopher Columbus brought back reports of the bizarre American vegetable from the hinterlands of Cuba. At that moment hundreds of varieties of maize, brimming with genes adapted to different habitats, carpeted the American continents. And an assortment of related wild grasses, not yet displaced from their native range by the invasion of uniform hybrid corn, urban development and other human activities, still languished as weeds along the perimeters of cornfields, where they occasionally swapped genes with adjacent maize plants.

These hardy local variants were nowhere near as productive as their modern descendants. But because each crop harbored a hodgepodge of different genes, neither were they as likely to be devastated by insects or disease. In the face of almost any environmental insult, there would always be plants possessing a mix of genes that would equip them to survive.

Laid side by side, ears from each of the six main types of corn seem to suggest that the *Zea mays* still retains a remarkable level of physical variability. And compared with wheat, rice and some other widely grown cereals, it does. But this remaining variability should not make us complacent.

The genetic resources of maize and many monoculture crops have been eroded during this century. Much of that loss can be attributed to the relentless destruction of habitat — through urban sprawl, an expanding monoculture and the clearing of primeval forests. In a world that is now witnessing the extinction of about a thousand species of plants and animals each year, the disappearance of little-known ancestors and a few weedy relatives of modern domestic species should not surprise us. But we may have to pay a price if we continue to squander these genetic resources in the years ahead.

The domestication of maize during the past five centuries has led to the triumphant harvests of the twentieth-century Green Revolution. But it has also brought about a steady loss of this precious pre-Columbian bounty of diversity in the germ plasm of maize. That trend can be traced in part to the very nature of domestication — the process of gradually adapting a plant or animal species to life in intimate association with, and to the advantage of, human beings.

The strategy behind this human-driven form of evolution is artificial selection — the systematic winnowing of hereditary traits we define as "good" from those we define as "bad." Its immediate aim is generally to increase the quality or productivity of crops and livestock; its long-term effect is usually to reduce the stores of genetic variability within a given species.

The Value of Genetic Diversity

From an evolutionary perspective, the maintenance of genetic diversity is a powerful strategy for a species' survival. Locked away in a species' germ cells — its total inventory of sperm, egg, seeds or other assorted reproductive cells that serve as a genetic bridge between successive generation of organisms — lies a reservoir of genetic options, or alleles, for each trait that together constitute the hereditary raw material of natural selection.

This resource of genetic variability allows a species to adjust, over time, to the endless perils of an ever-changing environment. It makes for a system capable of a great deal of trial and error.

The "trial," as discussed in chapter 4, takes place as living organisms are pitted against one another and against their surroundings. "Error" results when such an organism's genes, through death or failure to reproduce, are not transmitted to its offspring as successfully as are those of other members of the species.

In evolution's grand scheme, genes — singly or in combinations — can be rendered instantly obsolete by a quick shift in competition, climate or some other unpredictable factor. Yet these same genes might just as quickly prove useful to survival in another time or place. This means that systematically culling genes that might appear to be useless — for domestication or any other purpose — is not simply a form of hygienic housecleaning. It can also be a blind destruction of the potential options in a species' evolutionary future.

There is nothing unique about genetic diversity in maize. Evolution in all species — from hemlocks to *Homo sapiens* — relies on a ready reserve of alternative traits, or alleles. That does not mean that some kinds of artificially inbred, genetically similar creatures, like breeds of Siamese cats or laboratory mice, can't enjoy some measure of evolutionary success. But unless some remnant of that species' precious reserve of genetic variability is preserved, these animals may face future environmental demands genetically flat-footed — or, if released into the wild, lack the genetic resources to adapt to the fickle demands of a complex, eternally changing, natural ecosystem.

The story of maize vividly illustrates some of the possible consequences of relying too heavily on artifical selection in any species — human or nonhuman — and our historical tendency to pursue short-term gains without regard to their possible long-term consequences.

The Basis of Genetic Variability in Maize
By selectively sowing seeds from corn plants that were hardy, many-kerneled and tall, early agriculturists of the New World systematically discarded countless alternative genes in the maize

gene pool. But domestication is an inexact process and not all genes lost were harmful.

But the twentieth-century decline in the genetic diversity of maize populations is a loss of a different order. It is not simply the result of a relentless artificial selection for economically useful maize traits. It is also a consequence of the destruction of the very species — wild and domestic grasses — that are the genetic foundations of modern maize, along with the ecosystems that support them.

Despite the magnitude of that ecological damage, adequate funds are still not available for scientists to study and preserve the priceless genetic variability contained in the gene pools of living relatives of maize and other economically important plants. On occasion, scientists have tried to establish centralized seed banks in which kernels from a wide range of maize varieties are carefully maintained. Or they have attempted to set aside small sanctuaries to protect threatened native grass species. But such stop-gap measures are expensive, and it is widely agreed they are doomed to failure unless bold new international initiatives aimed at the conservation of these genetic resources are put into place.

Meanwhile, the evolutionary footprints of maize etched in the gene pools of local, genetically distinct varieties of corn that have been sustained by traditional farming techniques in some areas of the world and of wild weeds bearing little resemblance to their domesticated cousins continue to disappear. And replacing them are increasingly uniform maize crops designed not to protect the genetic diversity of *Zea mays* but to integrate it into an emerging global system of mechanized, chemically dependent, industrial agriculture.

Over the past several decades, the same remnants of genetic variability in the maize gene pool that have been so successfully exploited by agricultural researchers to breed high-yielding corn hybrid varieties have also endeared *Zea mays* to academic geneticists engaged in basic research. The reason is simple. To begin to track the shadowy trail of genes from one generation to the next, scientists needed an assortment of visible traits that could mark, like bright-colored flags, the movements of genes and chromosomes during carefully controlled test matings. Because its genes are so readily mapped, maize — along with fruit flies, bread molds and a few other unlikely laboratory organisms — has been transformed into one of the most intensively studied species in the history of genetics.

The closely watched matings of maize have, for instance, helped illuminate the processes by which individual genes sometimes carry out intricately coordinated activities within a cluster of genes, in much the same way that individual musicians interact to perform a complex piece of chamber music. Beginning in the 1940s, with the work of the pioneering American geneticist, and later Nobel laureate, Barbara McClintock, experiments with maize also led to the important discovery of eccentric, acrobatic "jumping genes," called *transposons*, that may one day be harnessed to transport new genes to future genetically engineered crops.

But the value placed on specific traits within a species' store of genetic variability is bound to differ considerably from one individual to the next, depending on one's motives. On the one hand the commercial plant breeder has learned to covet diversity in the form of genes that might be exploited by crossing plants with economically important hereditary traits — those affecting the percentage of protein in a corn kernel, for example, or the temperature tolerance of a newly sown seedling, or the ability of a growing plant to withstand periodic pesticide sprayings or botanical disease.

On the other hand, many research geneticists in search of broader biological themes have found delight in bizarre, often economically laughable traits. A walk through the test fields of such a scientist might, to the uninitiated, seem more like a stroll through a carnival sideshow of genetic freaks. Some plants might have cobs bearing flamboyant kernels of uniform scarlet, translucent white or jet black. Others might have leaves that instead of being familiar chlorophyll green, look pin-striped like a banker's suit — or sport ears tipped with corn silk colored eye-catching, shades of pink, not straw yellow. Yet, whether one is an applied researcher — dedicated to the search for profitable new agricultural spinoffs of genetic principles — or a basic researcher — concerned solely with adding to the body of knowledge — the biological resource is identical: the native genetic variability of maize.

Where are the roots of genetic variability in maize? And what natural mechanisms exist in this species to make the most of the genetic variability it still retains? The answers lie in a number of fundamental evolutionary processes described in chapter 4. The primary source of genetic variation in maize, as in all species, is the process of mutation — the sporadic molecular goofs and glitches that inevitably garble the coded messages of genes. A secondary source is meiotic cell division — the tightly choreographed two-act division of sex cells, described in chapter 3, that not only generates egg and sperm cells in sexually reproducing organisms such as maize but also serves to regularly reshuffle the random genetic changes created by mutation.

Sexual Reproduction in Maize

You may have never thought of corn as a plant with a sex life. In fact, it has a rather intriguing one (see Figure 12.3). Corn is hermaphroditic — a corn plant possesses both male and female reproductive organs. Unlike all other major grain species, it is equipped with male and female flowers located on separate parts of the plant.

The tip of a mature maize plant bears large numbers of tiny drab male flowers on a drooping pendant of tassels. Hundreds of small female flowers form distinctive cylindrical bouquets on each cob. Natural selection has seen to it that male and female flowers on the same plant generally do not mature at precisely the same time. As a result, corn plants usually do not pollinate themselves. This feature encourages genes to flow from one plant to another, stirring the gene pool, and thereby enhances overall genetic diversity.

Sexual reproduction in a field of maize takes place something like this. First, millions of pollen grains, shed from the masculine tassels, drift aimlessly on summer gusts of wind toward receptive plants nearby. When a grain happens to settle on the sticky blond tangle of corn silk that protrudes from the tip of an ear, it forms an intimate bond with a single silken thread. Each such thread, it turns out, is a microscopic tube attached, like a lifeline, to the ovary of just one female flower on the ear. It is through this living pipeline that the genetic cargo of one pollen grain passes. Eventually it releases a haploid male nucleus that fuses with its female counterpart inside the flower's ovule — or kernel-to-be. This sexual union yields a minute maize embryo, lying dormant in the food-rich sanctuary of a seed.

In addition to the continuing processes of mutation and meiosis, a legacy of genes from the past of maize contributes to the genetic diversity of maize. The accumulated evolutionary experience of tens of thousands of generations of maize plants is contained in the gene pool of modern maize. While these variant

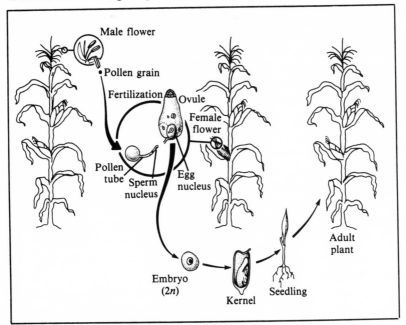

FIGURE 12.3: Life Cycle of Maize

forms of genes first arise too through transient mutational events and subsequent meiotic recombinations, they have also been sifted by centuries of natural selection.

Together these genes constitute a rich reservoir of genetic messages that have already passed the survival tests posed by earlier environmental conditions. Some of these may play a role in the adaptation of maize plants to contemporary conditions; others may drift idly in the maize gene pool for now, only to be drawn upon to help a plant face unforeseen environmental challenges. In the end, the gifts from all three sources of genetic variability in maize — mutation, meiosis and a legacy of variant genes from the species' evolutionary past — are united in the species' quota of reproductive cells. For it is this enduring lineage of germ plasm — the fragile thread of viable seed that connects each generation of maize — that harbors this priceless diversity of DNA sequences.

Individual genes or clusters of genes — the ultimate currency of genetic variability — in maize and its closest kin can suffer a number of possible fates. Some of these DNA sequences will disappear into the dark abyss of genetic death whenever a corn plant perishes, for whatever reason, before reproducing. They can drift passively — as hidden recessive alleles or as alleles that are scarcely expressed in the phenotype of a plant — within the vast collective current of genetic information that flows from one generation of maize crops to the next. Some achieve membership in the living pool of maize genes on their own merits; others gain admission simply because they happen to be linked to more influential genes on maize chromosomes. And it is always possible that on rare occasions a particular DNA sequence will surface unexpectedly to find expression in an exceptionally well adapted plant that will turn out to be a major innovation in the evolutionary history of the species.

Whatever their paths, the genes that have finally found their way into the gene pool of *Zea mays* at any instant in time represent, like the libraries of a great civilization, the collective, codified memory of a species.

Hybrid Corn

The innovation that enabled the rise of *Zea mays* to the exclusive club of global monocultural crops — genetically uniform plants whose seed is distributed commercially and sown in diverse regions of the world — was the breeding of hybrid corn. It was a brilliant strategy, brought about by a small group of American plant geneticists beginning in 1917.

It was not a radical departure from classical plant breeding. Both techniques achieved genetic exchange between parent strains largely by controlled matings that reshuffled genes and chromosomes between parental lineages. But the breeders of hybrid corn added a novel twist to this traditional strategy by deliberately reducing the level of genetic variability in parent plants before they were mated.

No single formula exists for breeding all types of hybrid corn. But the step-by-step recipe for creating one of the most common commercial corns by a double-hybrid cross is illustrative. The basic scheme is twofold: First mate, or hybridize, two sets of plants derived from two distinctly different genetic strains; then mate the freshly hybridized offspring of the two crosses. The seed of the final progeny is hybrid corn (see Figure 12.4).

All four of the founding parental strains (let's call them varieties A, B, C and D) have already been independently bred for a high degree of genetic uniformity. To do this, strain A plants have repeatedly been crossed with other strain A plants for a number of generations. Strain B plants have bred strictly with their immediate relatives. And so on.

Imposing such a strict schedule of inbreeding within a small group of genetically similar organisms causes each to experience a steady loss of genetic variability. In later generations, the sites of genes for many traits within strain A tend to have identical, or homozygous, alleles — as do those in each of the remaining three strains (though for different genes in each strain). As they become increasingly homogeneous, each generation of plants also tends to be smaller and more feeble and to yield a poorer harvest.

In the first of two hybrid crosses leading up to commercial

stocks of double-hybrid corn seed, the breeder crosses the relatively pure A and B lines to yield a hybrid plant (labeled AB). Reflecting its genetically dissimilar parents, it has at least two different alternative genes at almost every site along its chromosomes; we say that such a plant is highly heterozygous. A similar cross of homozygous C and D lines produces a second hybrid plant (CD) that is also highly heterozygous but bears a different array of variant genes.

In the next cross, the breeder mates these two hybrid progeny

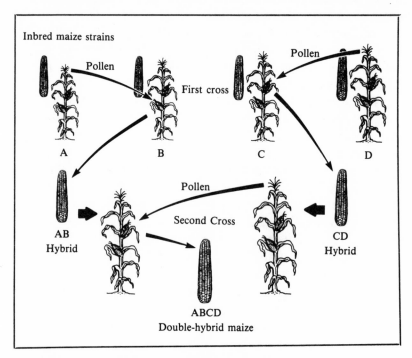

FIGURE 12.4: Maize Double-Hybrid Cross

— the heterozygous AB and CD plants — to complete the double-hybrid game plan. By thus combining two generations of strains whose genetic differences were artificially exaggerated, the resulting offspring is gifted with a remarkable number of genetically different, or heterozygous, alleles. The immediate advantages of this new hybrid plant become apparent soon after its seeds are sown. Its yields are likely to far surpass those of any of its immediate ancestors, not to mention those of other traditional varieties of maize.

Heterosis in Hybrid Corn

The key to the success of hybrid corn lies in a still somewhat murky concept called *heterosis* — the general tendency for the hybrid progeny of most species to be more robust than their inbred parents. It is common knowledge among ranchers, for instance, that highly inbred breeds of cattle must occasionally be crossed with related breeds to maintain the strength of the line. But the tonic effects of this sudden infusion of new genes has never been fully explained. In most cases, they can be attributed to the explosive appearance of heterozygosity — alternative alleles at a large number of gene sites. This meeting of unmatched genes seems to generate a flurry of structural changes in many enzymes — some of which fuel plant metabolic rates, seed development and growth.

But heterosis in hybrid corn also has a price. Farmers who purchase such hybrid seed and sow it, along with the requisite doses of chemical pesticides and fertilizers, are indeed rewarded with more productive crops. But the sudden surge of hybrid vigor is short-lived. If farmers attempted to set aside a portion of their increased harvest for next season's seed, they would be sorely disappointed. The exceptional vigor associated with hybrid seed is generally lost within a few generations because inbreeding using homegrown seed tends to reduce genetic variability and yield smaller and less productive crops.

To avoid this problem, farmers are forced to return to the genetic alchemists who bred the first sack of double-hybrid seed. These commercial plant breeders, fully aware of the limits of hybrid crops, have meanwhile been busy concocting a new generation of rejuvenated hybrid seed for next season's planting. Thus, farmers who have benefited by hybrid seed have also simultaneously become economically "addicted" to their corporate sources of seed.

The Genetic Vulnerability of Hybrid Corn

In 1970, an epidemic of Southern leaf blight fungus in the United States dramatized a second possible hidden cost of cultivating hybrid corn — its inherited vulnerability to disease. Beginning in Florida, the outbreak of Southern leaf blight was a wave of infections in corn crops caused by a species of pathogenic mold — one of the hundreds of species of bacteria, fungi, viruses and insects that are the natural adversaries of maize.

The first symptoms of the disease were rather subtle. First, the leaves and stems of corn crops became speckled with tiny brown diamond-shaped spots. Soon the small patches of dead cells spread until plants were streaked with brown. Many of them died. The outbreak spread like wildfire, traveling at speeds of up to 80 kilometers a day. From northern Florida it swept eastward through several other southern states and then veered northward, aided by favorable weather conditions, into the central Corn Belt states that form the heart of American corn production.

What was striking about this raging epidemic of Southern leaf blight was not that the mold could kill corn crops. The species was already a well-known, but minor, local pest that destroyed an average of less than 1 percent of the annual U.S. harvest. It was the way the mold suddenly seemed able to cut a large swath through cornfields over an enormous geographic area. In a single season, the powdery mold — a species named *Helminthosporium maydis* — had become a national scourge. By the end of the 1970 growing season, Southern leaf blight had devastated more than 12 percent of the annual U.S. harvest and more than the anticipated losses from all other corn diseases combined.

Why had American hybrid corn suddenly succumbed to what appeared to be just another old adversary? One reason lay in the very essence of hybrid corn's success — the assembly-line uniformity of its commercially bred seed. The procedures for producing commercial hybrid corn seed guarantee a certain amount of uniformity from seed to seed. It is simply part of the price of pursuing large and predictable harvests. The bulk of this uniformity is genetic — a repetition of a limited number of commercially important genes throughout much of the North American range of maize.

In recent years, one could argue, our ancient symbiotic bond with maize has grown strangely warped and distorted. Through ingenious breeding techniques, we have managed to spur our crops to greater productivity. But at the same time we have become increasingly reliant on industrial chemicals to mask their evolutionary vulnerabilities and flaws. In the midst of our astonishing agricultural successes, we are surrounded by the persistent residues of pesticides in soils, foods, water supplies and even our bodies.

Not every hybrid corn plant, of course, is the genetic twin of every other plant. There are many varieties of hybrid corn, and each varies genetically from its hybrid cousins. But there does exist a degree of genetic uniformity at certain critical gene sites along maize chromosomes. If these shared genes happen to code for the plant's genetic resistance to a specific disease such as Southern leaf blight, they can create a distinct vulnerability — a hereditary Achilles' heel — in corn. For such rigid, homogeneous, hereditary barricades to infection can readily be overrun by genetically agile pathogens equipped with their own precious reserves of genetic variability.

This is precisely what occurred during the infamous Southern leaf blight epidemic. Later studies showed that an estimated 80 percent of the 1970 U.S. corn crop possessed a heritable trait known as male sterility factor, which plant geneticists had intentionally bred into hybrid corn crops. The reason the male sterility trait was introduced into hybrid corn populations was to genetically emasculate plants so that they would not accidentally fertilize their own flowers. But unbeknownst to well-meaning plant geneticists, another, temporarily hidden trait turned out to be closely associated with male sterility factor — an inherited susceptibility to the poisons secreted by the fungus that causes Southern leaf blight in corn.

Oddly enough, the genes responsible for the male sterility factor in hybrid corn were most atypical ones; they were not transmitted to progeny on parental chromosomes. Instead, they resided, and replicated, independently of mainstream genes — on tiny plasmid particles of DNA in the energy-producing mitochondria of certain reproductive cells in maize. Unlike most genes in the cell, these maverick extrachromosomal brethren are transmitted to progeny as flotsam in the fluid cytoplasm of the maternal egg cell lodged in each kernel of corn. Since such cytoplasmic juices, as well as genes, are part of each egg's gift to the next generation, these genes can be inherited only from a maternal reproductive cell. But, like any repetitive genes, they too add to the cumulative burden of genetic uniformity in maize.

Genetic Engineering and Genetic Diversity

In the years ahead, efforts to genetically engineer novel domestic plants could make it even more difficult to conserve the genetic resources of our crops. For the moment, the art of agricultural genetic engineering is still in its infancy. Shuttling genes between plants has so far proved to be far more difficult than it has been using some microbial or animal species. But assuming our gene-splicing skills improve, one cannot help but wonder what frightful new burden of genetic uniformity may be in store for future generations of farmers.

It may come, for example, in the form of commercially successful recombinant crops — armed with identical batteries of foreign genes coding for resistance to herbicides, drought and disease — or as plants and seeds nurtured from laboratory tissue culture created from a single recombinant cell. These cloned crops would share not just a few genes in common like modern hybrid corn plants. Their synthetic genomes could be virtual photocopies of each other — identical at every gene site.

In such cases, the genetic engineering of agricultural crops could further exaggerate the existing genetic uniformity of crops. It could open the way to new levels of genetic monoculture that might render crops even more vulnerable to changes in climate, soil conditions and disease than they already are. At the same time, the short-term success of genetically engineered crops could further displace the last remaining populations of some of the ancestors or relatives of our modern domestic plants, thereby further eroding our genetic resource base.

It could also be legitimately argued that some recombinant DNA techniques promise, by exchanging genes between species that do not normally interbreed in nature, to enormously enrich the genetic diversity of some species. But genetic diversity, it must be remembered, cannot be measured simply by the quantity of genes available to a species. Genetic diversity also includes a qualitative dimension. For one must be willing to assign a special value to those genes that not only code for a desirable trait but also have been subjected to generations of evolutionary tests and have proved able to contribute to a species' survival while working in harmony with tens of thousands of fellow genes.

The Moral of the Maize Story

The Southern leaf blight epidemic is a reminder, too, of another easily forgotten evolutionary truth. All species — not just edible ones like maize — are in a continuous state of genetic flux. The same forces of natural selection that shape human beings and maize plants also shape molds. The outbreak of Southern leaf blight resulted not from the familiar form of mold, as most had at first suspected; it arose from a newly evolved strain that may have differed from its gentler cousin by only a few genes. This brief burst of novelty in a mold allowed it to quickly pierce the unified armor of genetic defenses that surrounded hybrid corn crops.

The natural enemies of plants have their own stores of genetic diversity. And we are simply not knowledgeable enough to anticipate what their next clever adaptation will be. As a result, we tend to rely increasingly on chemical pesticides to shield our precious domesticates from the wild. As keepers of maize, we are now responsible for continuously propping up its crippled genetic defenses. We can only hope to do that if we come to look on each species' store of genetic diversity with profound respect.

For maize and other domestic species, that means taking steps to preserve the habitat and germ cells that harbor what little genetic variability remains. There have been some efforts, to be sure, in this regard. A few natural sanctuaries have been established where native populations of important plant species can continue to thrive in their natural habitat. Some countries have created special seed banks to preserve precious germ plasm reserves — in the form of viable seed and fertile plants. Because of the strictly controlled environmental conditions required to ensure the viability of seed, however, such programs can prove prohibitively expensive for many developing countries. Some scientists continue to doggedly identify and catalog potentially useful relatives of domestic plants from surviving stands of native crops and weeds in natural centers of diversity within the Third World. And some farsighted organizations continue to promote more enlightened agricultural strategies that would reward greater regional reliance on genetically diverse crops rather than an unthinking dependency on monoculture.

But without dramatically increased levels of international cooperation and funding for the conservation of genetic resources in agriculture, many believe that such efforts are doomed to failure. In the process, we will have squandered an irreplaceable genetic resource. Maize's original rich store of genetic variability could not possibly arise again from chance mutations, meiotic recombination or even clever surgical transplants from other species using emerging recombinant DNA techniques. For we will have lost much of the species' precious legacy of genes, including many that evolved together in units and long ago demonstrated their evolutionary worth. Some of these lost genes could conceivably have been adapted to environmental challenges that maize faces today — not to mention future, unknowable challenges that the species has yet to overcome.

Knowing this, we can look out again upon our imaginary field of sunstruck Iowa corn and see in it more than a symbol of our mastery over another species' fate. Without diminishing the gifts modern agriculture has granted us, we can also see in this scene evidence of our own limited vision of the complexities of evolution. We naively set out to shape species to meet our needs and are for a time amply rewarded. But there are moments when we are painfully reminded of the dimensions of our ignorance. At such times, we are understandably overwhelmed by the enormity of our chosen task — trying to anticipate the simultaneous evolutionary twists and turns of tens of thousands of species, not just men and maize.

CHAPTER 13

MAPS AND DREAMS:
DECIPHERING THE HUMAN GENOME

GENETHIC PRINCIPLE
The accumulation of genetic knowledge alone — however precious in its own right — does not guarantee wisdom in our decisions regarding human heredity; if such knowledge breeds a false sense of human mastery over genes, it can even lead to folly.

The total human sequence is the grail of human genetics.
— Walter Gilbert, in *Science*

Many things in the world have not been named; and many things, even if they have been named, have never been described.
— Susan Sontag, in *Against Interpretation*

The stupidity of people comes from having an answer for everything. The wisdom of the novel comes from having a question for everything. . . . The novelist teaches the reader to comprehend the world as a question. There is wisdom and tolerance in that attitude.
— Milan Kundera, in *The Book of Laughter and Forgetting*

The Geography of the Human Genome

In its physical dimensions, the human genome — the entire array of gene-bearing DNA sequences contained in a human cell — is incredibly small. Even if all 23 glistening threads of DNA in a human egg or sperm were spliced end-to-end and stretched into a single, taut nucleic acid molecule, it would measure a little over one meter in length. Yet this fine thread of hereditary material would contain a mind-boggling quantity of genetic information. For within the spiraling central core of this single, fused, double-helix molecule are approximately three billion nucleotide base pairs — a linear assortment of adenines, thymines, cytosines and guanines. In a living human reproductive cell, this landscape can be seen as a miniature archipelago of nucleic acids — 23 chromosomal islands of DNA, each harboring a portion of the genome's estimated 100,000 genes.

Historically, one of the greatest challenges faced by modern genetics has been the attempt to map the precise location of these genes on the 23 chromosomes in this microscopic genetic realm. In this sense, geneticists can be seen not only as explorers but also as cartographers of human genetic material — laboriously charting the wilderness of the human genome. The notion of the "geography" of genetic material, in fact, is one of the guiding metaphors of modern genetics and lies at the heart of some of the most vivid scientific terminology used to describe the structure and organization of the human genome. The most visible features of human chromosomes — their centromeres, arms or characteristic staining regions, for instance — are considered "landmarks" in the genetic terrain. Distances between genes are paced off in such minuscule units of length as the *kilobase* (a length of DNA made up of 1,000 nucleotide bases) or the *centimorgan* (an indirect measure of the distances between genes based on genetic crossing over during meiosis). Individual human gene sites, chromosomal regions and entire chromosomes are even assigned unique names — like so many lunar craters and ridges in a newly explored moonscape.

To date, geneticists have succeeded in painstakingly plotting the geographic locations of more than 4,000 genes in the human genome — some more precisely than others. That represents only about 4 percent of the total number of protein-coding, or structural, genes in your body. But only in the past decade or so, with the emergence of powerful recombinant DNA techniques and other related genetic technologies that permit the cloning and rapid sequencing of human DNA molecules, have scientists begun to believe that the geneticist's ultimate dream — the complete accounting of every chemical base in the human genome — might be realized during their lifetimes. Today many experts are predicting that we might possess a complete record of the genetic content of *Homo sapiens* by the year 2000.

The significance of this chemical inventory of human DNA is not yet fully known. But many geneticists are convinced that a complete base-by-base sequence of the human genome could at the very least be used as a road map by future researchers attempting to locate and classify human genes — from medically important genes associated with cancers and hereditary diseases to genes regulating fundamental metabolic processes in human cells. They also believe that by revealing the chemical basis of a wide range of subtle hereditary differences between individuals they will advance our understanding of human biology on many fronts — from physiology and disease to comparative evolution. In the process, the sequencing of the human genome is bound to provide precious clues to the organization and function of the human genetic apparatus.

Early Gene-mapping Strategies

Historically, one of the basic methods used to determine the chromosomal location of genes has been to carry out controlled-breeding experiments with laboratory organisms — ranging from garden peas to fruit flies — possessing known genotypes, and then to follow the fate of specific hereditary traits over many generations.

During the first decades of this century, such controlled matings led to the discovery of a phenomenon known as gene linkage — the tendency of neighboring, nonallelic genes located on the same chromosome to remain adjacent during meiotic cell divisions and, therefore, to be inherited together in organisms of the next generation. As the distance separating two such genes increases, so does the likelihood that during meiotic crossing over the two genes will be physically separated and travel along separate hereditary paths to different reproductive cells and from there to different offspring.

Knowing this, classical geneticists were able to use experimentally determined crossover frequencies of hereditary traits to sketch crude linkage maps showing the linear, beadlike order of individual genes on chromosomes. Unfortunately, because distance in these maps was determined indirectly — from meiotic recombination rates that varied considerably from one species to the next — its accuracy varied considerably depending on which laboratory organism one was studying. Thus, such maps only revealed relative spatial relationships between genes rather than actual physical distances.

Family Pedigrees
These first maps of genes were based on controlled matings between organisms of known genotype. Because human breeding experiments are ethically unacceptable, geneticists must rely on other methods for evidence of human gene linkages. One of the oldest methods is the study of family histories, or pedigrees. By laboriously recording the appearance of a particular trait over several generations of a family tree, it is often possible — even with a relatively small sample of individual genotypes — to detect patterns of inheritance that would have quickly been revealed by deliberate breeding experiments. If these patterns of inheritance can then be correlated with patterns of chromosomal transmission, it is sometimes possible to determine the precise chromosomal address of a gene.

As early as 1911, geneticists used the family pedigrees to assign the location of the human gene responsible for red-green color blindness — a hereditary condition that disrupts normal color vision in males — to a specific chromosome. Their task was enormously simplified by the fact that the color-blindness gene resides on one of the two sex chromosomes. As you may recall, the X and Y sex chromosomes determine hereditary gender — XX humans are females; XY humans are males. Thus, simply by knowing the sex of an individual, geneticists know something about that person's chromosomal constitution. When family studies revealed that the color-blindness gene was typically transmitted from mother to son, not from father to son, it became apparent that the gene was linked to the X chromosome (see Figure 13.1). In a similar fashion, geneticists have over the years managed to assign more than 200 other human genes to the X chromosome. In addition, a few genes have also been located on the much smaller Y sex chromosome.

Chromosomal Banding Patterns
But the vast majority of human genes reside on autosomes — the remaining 22 pairs of non–sex chromosomes. And unlike the sex chromosomes, autosomes are transmitted independently of gender. Thus, the sex of an individual cannot, alone, reveal the presence of a particular chromosome. To associate a specific hereditary trait in a family pedigree with a single non–sex chromosome, geneticists needed methods to distinguish one autosome from another.

In the 1950s, increasingly sophisticated techniques were developed to display the full complement of 46 chromosomes in a human body cell by arranging them in a karyotype — an orderly, microscopic lineup of chromosome pairs on a slide. For the first time, human geneticists could begin to make detailed comparisons between individual chromosomes and recognize each chromosome by its structural markings.

In 1969, geneticists discovered that they could artificially highlight subtle chemical differences in chromosomal anatomy by dyeing chromosomes with Giemsa or fluorescent quinacrine mustard stains. Dyeing chromosomes had a remarkable effect:

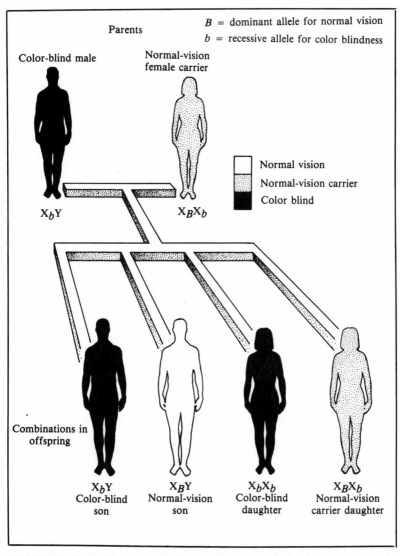

FIGURE 13.1: Human Pedigree Showing Inheritance of Red-Green Color Blindness

It revealed a multitude of parallel, zebralike stripes running perpendicular to the length of the chromosome. Each chromosome, it turned out, possessed a unique pattern of horizontal bands. Because the banding patterns of each of the 23 pairs of homologous chromosomes usually vary slightly from one person to the next, they can be used as chromosomal finger-prints to identify individual chromosomes (see Figure 3.2). Only later was it learned that the stained bands corresponded to regional differences in the density of cytosine-guanine (C-G) and adenine-thymine (A-T) base pairs within the single convoluted DNA molecule comprising each chromosome.

Besides establishing the individuality of each chromosome, the pattern of bands had another useful function. It could be used as a relatively stable grid of chromosomal landmarks — called markers — in the human genome. There were, to be sure, already a number of highly visible landmarks available — the darkly stain-ing regions of quiescent DNA on chromosomes known as heterochromatin, for example, or the knotlike centromeres by which they are dragged during mitosis and meiosis. But the sheer number of stained bands granted geneticists a higher-resolution image of the chromosomal landscape.

When, during the course of an imperfect cell division, a chromosomal fragment chanced to be lost, duplicated or ex-changed, that alteration might be made visible as a disruption, or "scar," in a chromosome's usual banding pattern. For example, individuals suffering from retinoblastoma — a poten-tially fatal hereditary form of eye cancer that results in tumors of the retina — can sometimes be shown to have segments miss-ing from a particular chromosome. This deletion — visible through a microscope as a minute gap in that chromosome's normal banding pattern — suggests that one gene implicated in this illness can at least be tentatively assigned to that chromosome. As banding techniques grow more sophisticated their value in identifying structural changes in human chromosomes associated with certain genetic disorders will also increase.

Somatic-Cell Hybridization
Despite the ingenuity of such schemes, progress in human gene mapping proceeded at a glacial pace until the late 1960s. At that

time, a radically different approach to locating individual genes on chromosomes was developed. In *somatic-cell hybridization*, human cells and mouse cells are artificially fused to form hybrid cells. Because these chromosomally bloated cells have a habit of ejecting human chromosomes and retaining mouse chromosomes, researchers can carry out laboratory experiments designed to reveal the activity of a single gene in an arena of only a few human chromosomes — instead of the full complement of 46 chromosomes. Thus, if locating a gene in the human genome is like looking for a needle in a haystack, somatic-cell hybridization techniques reduce the size of the genetic haystack.

This is how somatic-cell hybridization works (see Figure 13.2). A culture containing a mixture of human cells and mouse cells is deliberately infected with a specially prepared microbe called a Sendai virus. Because the outer surface of this virus is bristling with molecular receptor sites that allow it to bind to a host cell, the virus can adhere to more than one cell at a time. As it grips neighboring human and mouse cells in a tight embrace, the cell membranes of the two cells sometimes fuse, forming a hybrid human-mouse cell possessing two complete sets of chromosomes in its nucleus — one from each parent species. During cell division, these genetically overburdened cells tend to retain their mouse chromosomes but gradually jettison human chromosomes.

Colonies descended from these unnatural hybrids, each laden with only a few remnant human chromosomes, are then grown on nutrients that can sustain only those cells harboring a specific human gene. Because cells lacking the critical gene are unable to metabolize the nutrients available in the restrictive medium, they starve to death. By comparing the chromosomal contents of the survivors, scientists can determine which one of the original 46 human chromosomes occurs in every cell. Because each survivor must harbor a copy of the target gene, the appearance of a particular chromosome in every cell confirms that the gene is located on that chromosome.

In this manner, it has been possible to assign dozens of human genes to specific chromosomes. Compared with the painstaking process of tracking genes through family pedigrees, somatic-cell

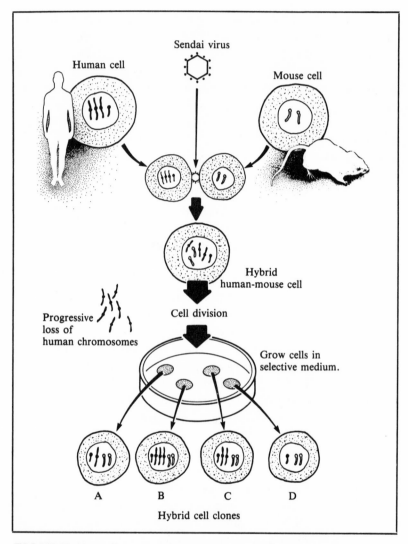

FIGURE 13.2: Somatic Cell Hybridization Technique

hybridization is remarkably quick and effective. But it also has serious limitations. Only genes responsible for hereditary traits that can be readily detected in the laboratory — by selective nutrients or other methods — are suitable for the cell hybridization technique. Because most human genes do not meet this requirement and cells harboring them cannot easily be distinguished from cells that do not, efforts are being made to map them using other strategies.

Recombinant DNA and Related Techniques for Mapping Genes

The most powerful of these new strategies depend on recent advances in molecular genetics and recombinant DNA technology. These tools grant geneticists an enormously magnified view of the human genetic landscape — one that can reveal not only the chromosomal location of a gene but also its ordered sequence of nucleotide bases. Once researchers have determined the precise chemical identity of a gene, that base sequence can be cloned into abundance in bacteria cells for closer study or, in some cases, used to infer the amino acid formula of an unknown polypeptide product of the gene.

For instance, using gene-splicing techniques, geneticists can clone small, single-stranded segments of radioactively labeled DNA or RNA for use as mobile gene-mapping probes. When the radioactive probe, derived from a gene product or the nucleic acids used to build it, is mixed with the unraveled DNA strands of human chromosomes, it homes in on and bonds with its complementary sequences. Its radioactive emissions then serve as a beacon marking that gene's position in the genome.

Techniques can also be used to dissect human chromosomes into smaller pieces in order to examine them more closely. To do this, human DNA molecules are broken up into smaller pieces using restriction enzymes that cleave DNA whenever a specific base sequence occurs. These DNA fragments can then be scrutinized for clues that reveal the presence of a particular gene. For example, a radioactive DNA probe can be introduced to home in directly on that gene and reveal its location on one of the small restriction-enzyme fragments.

It is also possible to derive indirect clues to the site of a particular gene by measuring the lengths of the restriction-enzyme fragments themselves. If the DNA from a genetically "normal" person is shattered by a particular restriction enzyme, a set of DNA fragments of predictable length results. But a mutation that alters even a single base pair can eliminate an existing enzyme recognition site or create a new one. Thus, the application of the same restriction enzyme to the DNA of a genetically "abnormal" individual often results in one or more DNA fragments of distinctly different length. If such differences can be correlated with the appearance of a specific hereditary trait in the "abnormal" individual, that DNA fragment can sometimes be used as a genetic marker to monitor the inheritance of the still-hidden gene.

Restriction enzymes can also be used to produce rough, preliminary base-sequence sketches of a genome. Because each enzyme recognizes only a specific sequence of bases, each application of a new restriction enzyme highlights the presence of that sequence in the DNA sample under study. By systematically exposing the DNA sample to a succession of different restriction enzymes and combining those results, one can piece together a partial base-sequence map consisting of those recognition sites. Restriction-enzyme maps can provide only a fleeting glimpse of the total base sequence of a genome. But by identifying even a relatively small proportion of the total DNA sequence, they are capable of analyzing the chemical topography of the human genome in far greater detail than any of the other traditional gene-mapping techniques.

No modern technique is more impressive at charting the chemical contours of the human genome than rapid DNA sequencing. Its aim is straightforward — to decipher the messages of human chromosomes into an unambiguous list of three billion nucleotide bases. In this light, a gene no longer has the appearance of a kinetic molecular dancer. Instead it becomes a motionless bulletin board of nucleotide bases. Severed from their dynamic role in the hereditary and metabolic processes of a living cell, genes are reduced to their chemical essence — long strings of adenine, thymine, cytosine and guanine bases. Once sequenced, the incredible, multidimensional complexity of genes is reduced to rows of letters — A's and T's and C's and G's — that could be recorded by a child armed with nothing more than paper and crayon. Yet, in that same four-letter alphabet of nucleotide bases are inscribed the genetic blueprints for much of what defines us as human beings — from the tingling neurons of the brain's convoluted cortex to the fingerprint whorls of our opposable thumbs.

A variety of sequencing methods are available for directly analyzing the linear order of nucleotide bases in pieces of a human DNA molecule (see chapter 5). In each case, sequencing a DNA molecule depends on some sort of chemical sleight of hand to generate a series of single-stranded pieces of DNA — each one base longer than the last. These segments of DNA — each representing a part of one complete DNA sequence — can then be arranged according to length. Then, by identifying which base has been added to each successively longer chain of bases, one can gradually determine the linear sequence of bases in the entire DNA sequence. In this way, it is possible to systematically chip away at the coded secrets of human chromosomes without any knowledge of the role they perform in controlling or contributing to vital cell processes.

No other gene-mapping tool has aroused more enthusiasm among scientists in recent years than rapid DNA sequencing. This is largely because it makes the dream of completely deciphering the human genome technically feasible. Unlike some alternative gene-mapping methods that depend on knowledge of the activities or products of human genes, rapid DNA sequencing is capable of documenting the precise base-pair text of human genetic material long before geneticists have grasped the subtle interplay and output of particular human genes. In this sense, rapid DNA sequencing is like a farmer's combine poised at the edge of an enormous field of grain. The technology is already in place and ready to begin harvesting the vast store of genetic information that is chemically encoded in our chromosomes. Virtually all that is required to realize this stunning scientific achievement is a major commitment by governments and scientists of time, money and human resources.

The Grail of Human Genetics

One indication of how close we are to achieving that feat is the current debate, centered in the United States, about the merits of launching an all-out effort to sequence the human genome by the year 2000. Known as the Human Genome Initiative (HGI), the scope of the proposed megaproject would be unparalleled in the history of the life sciences. It is estimated that to sequence the three billion bases in the human genome, using existing genetic technologies, HGI could require up to 30,000 person-years of labor, the creation of huge centralized supercomputer databases to store sequences, and a budget exceeding $2 billion (U.S.). This requirement would place it among the ranks of such ambitious, goal-oriented Big Science projects of the past as the building of the first atomic bomb or sending astronauts to the moon — or such future projects as the American plan to construct the $4.5 billion Superconducting Super Collider particle accelerator or an $8 billion orbiting space station.

For the first time, we now have the opportunity to invest in a comprehensive base-pair map of our species' genetic makeup. The total base sequence of the human genome is there for the taking — for anyone willing to make the required investment. According to 1986 estimates, an industrious scientist, working alone and using current technology, could sequence about 100,000 bases — less than 1/10,000 of the total genome — in a year. The cost of that effort comes to about one U.S. dollar per nucleotide base — or about a thousand dollars for a typical human gene. That figure is certain to decline in the years ahead as faster, more efficient DNA sequencing devices enter the marketplace. Whatever the final price tag for sequencing the human genome, it is likely to be minuscule compared with the sums governments have allotted to the unending scientific search for more powerful military weapons.

Benefits — Applied Genetics

The benefits of this knowledge are likely to be enormous. For example, there will almost certainly be valuable medical spinoffs from a DNA sequencing project on the scale of HGI. As scientists methodically sequence their way through the human genome, they are bound to unravel a number of the DNA sequences of genes that either play a critical role in some human hereditary diseases or are linked closely enough to them to serve as useful markers. In this way, gene sequencing is likely to lead to all sorts of new diagnostic tests that can be used to recognize genetic disorders at an early stage, identify asymptomatic carriers of defective genes and provide improved genetic counseling for individuals and families. Newly sequenced genes will also be cloned in bacteria or yeast cells to churn out pure, genetically correct pharmaceutical products — ranging from human hormones and rare blood factors to brain neurotransmitters — that may provide new treatments for certain hereditary illnesses. In principle, these same DNA sequences could even play a major role in future efforts to actually cure genetic disorders by attempting to replace, modify or repair the defective DNA sequences in patients.

At the same time, we can also expect that gene sequencing will illuminate a vast array of subtle hereditary differences between individual human beings that are simply not detectable by any other means. Geneticists now believe that an average of 99.9 percent of all base sequences in human chromosomes are common to all members of our species. This means that only about three million bases in the human genome are responsible for the biological individuality of each person.

According to one 1987 estimate, a comparison of DNA sequences in any two human beings is likely to reveal variations involving about 1 nucleotide base out every 200 to 500 base pairs. As we continue to chemically map this hereditary terrain in the years ahead, we can expect to uncover many more individual differences in human DNA. The vast majority of these variant DNA sequences, or DNA polymorphisms, will probably turn out to be medically insignificant. Because more than one codon, or base triplet, can specify the same amino acid, people may possess slight differences in their DNA but produce identical protein products. For the most part, these variant DNA sequences will reflect little more than evolutionary footprints in the human genome — the accumulation of tens of thousands of generations of minute, mutational changes in our genetic molecules. Among them, though, will be some variant genes that will be shown to play important roles in determining our relative vulnerability to many human diseases. Thus, the description of these DNA polymorphisms could provide valuable clues for early diagnosis and treatment of illnesses ranging from lung cancer to schizophrenia.

In addition, as our list of known human DNA polymorphisms grows longer, geneticists will increasingly use these recognizable DNA sequences as molecular landmarks in tracking entire groups of human genes simultaneously through a family pedigree. Instead of trying to monitor the appearance of recognizable phenotypic traits in each generation, geneticists will be able to monitor the passage of certain hereditary traits by directly analyzing chromosomal DNA taken from blood samples of family members. In fact, this strategy has already led to a number of dramatic successes. Among them: the chromosomal location of the genes responsible for a hereditary form of Alzheimer's disease, an increasingly common degenerative brain disorder among the aged; for Huntington's disease, a devastating, late-onset neural disorder; and other conditions such as manic depression and cystic fibrosis.

Benefits — Basic Genetics

The most lasting rewards of sequencing human DNA may lie in the realm of ideas. Mapping the base-pair topography of the human genome will help refine our vision of the organization of the genetic apparatus of *Homo sapiens* and other higher organisms. In the same way that knowledge of human anatomy illuminates underlying physiological processes of the body, a better understanding of the molecular structure of the human genome is bound to offer unexpected insights into the mechanisms of genetic mutation and biological inheritance.

By analyzing mounds of raw DNA sequencing data, geneticists should be able to uncover important clues to the structure of individual human genes and the interactions between them. They should be able to recognize recurring patterns of control sequences adjacent to genes that will help explain how the genes in living cells are switched off and on during growth and development with such exquisite timing. And by comparing the DNA content of healthy and ill individuals, they may be able to monitor the passage of highly mobile, pathogenic DNA sequences, including retroviral genes that are relics of previous viral infections, as well as other categories of restless DNA such as oncogenes. Perhaps most important of all, a complete inventory of human DNA sequences will help future researchers to pose far more meaningful questions about the many mysteries of human genetics that will undoubtedly remain.

There is already compelling evidence of the power of DNA sequencing and related techniques to cast light on the shadowy molecular world of gene activity. Only by chemically matching the base sequences of chromosomal DNA in higher organisms with the sequences of messenger RNA molecules, for instance, did geneticists first realize that the transcription of RNA from DNA is an extraordinarily selective process.

Geneticists now know that the DNA molecules of eukaryotes — creatures, like ourselves, possessing chromosomes encased in a nuclear membrane — are riddled with DNA sequences that are never coded into protein. In fact, only a fraction of the information stored in the genome is faithfully transmitted to the metabolic machinery of the cell. Genes themselves, we now know, are strings of both coding and noncoding base sequences. They contain surpluses of base sequences — introns — which are enzymatically snipped out during the process of DNA transcription, leaving a lean, mobile final edition of messenger RNA that bears only those sequences — exons — that are essential to the final assembly of amino acids into protein.

DNA sequencing has also helped to highlight the astonishing degree of repetition in human genetic information. Some of the most common human genes are represented more than once in a cell's quota of DNA. For example, multiple copies of genes often provide sudden surges of RNA to ribosomes, the protein factories of cells. Redundancy, in this case, increases the output of a vital gene. But the human genome is also sprinkled with a perplexing variety of base-sequence repetitions — some present in tens of thousands of copies — that are too short to be typical genes. For the most part, their meaning remains obscure.

Benefits — Understanding Human Evolution

Sequencing the entire human genome will also help clarify the evolutionary history of our species. It will provide evolutionary biologists with the raw material for making detailed comparisons between the naked nucleic acids of different species with an accuracy never before possible. In time, as the total genetic content of other species is deciphered, evolutionary relationships between organisms will be established with greater precision. DNA sequencing will enable scientists to track the number and nature of genetic mutations accumulated by a species following its divergence from a shared ancestor, and this tracking will provide another useful "molecular clock" for measuring the passage of evolutionary time.

Again, a combination of preliminary base-sequence data and other, hazier, indirect evidence already provides us with hints of the sorts of insights such knowledge may bring. For example, DNA hybridization experiments, in which single-stranded human DNA molecules are allowed to chemically bond with DNA from a second mammalian species to form a hybrid double helix, have revealed a striking evolutionary kinship, or homology, between base sequences.

The DNA molecules of our closest living evolutionary family of relatives — chimpanzees and gorillas — are different from our own, it turns out, in only about 1 percent of their nucleotide bases. This means that about 99 percent of the chemical information stored in your DNA is indistinguishable from that of an ape staring at you in a zoo (see Figure 13.3). Common sense suggests that a variety of obvious physical and mental differences between humans and apes would contradict such findings. But the explanation for this apparent paradox seems to lie in a relatively small population of regulatory genes — those governing the pace and pattern of development — that exert an influence out of proportion to their numbers. In this sense, the human genome can be seen as an ancient primate genome that has been "humanized" over a period of hundreds of thousands of years by the appearance of a minority of critical "control" genes.

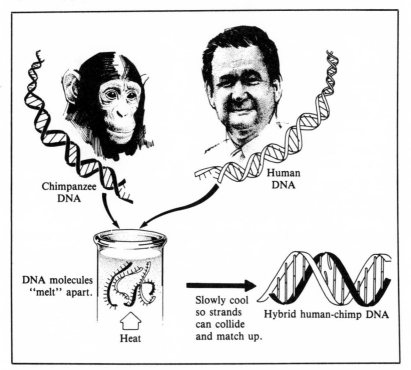

FIGURE 13.3: Human-Chimpanzee DNA Hybridization Experiment

Other interspecies DNA hybridization tests have underscored the close affinity between human and ape DNA. For example, two species of mice belonging to the same genus have been shown to share about 95 percent of the base sequences in their DNA. Despite the fact that the previously described hybridization between human and ape DNA bridged the broader evolutionary chasm between taxonomic families rather than species, it appears that the two visibly similar rodent species have approximately five times less in common, by this measure at least, than a Canadian prime minister and a playful chimpanzee.

The Possible Price of Sequencing Human DNA

An exaggerated sense of urgency sometimes surrounds discussions of an all-out campaign to sequence the human genome. In part, this urgency can be attributed to the same sort of insatiable curiosity that drives all scientists and to an eagerness to share in the rush of fascinating new facts about the structure and organization of the human genetic apparatus that have remained tantalizingly beyond the reach of researchers for years. It probably also arises from a feeling that a project to decipher the total human DNA sequence — the "holy grail" of modern genetics — seems to represent one of those rare milestones in modern genetics that could be reached with certainty during the lifetimes of most living scientists. Understandably, the thought of having a chance to help sift through the base-pair middens of the human genome within a decade or two — rather than bequeathing that task to posterity — must seem like a personal dream come true to many researchers.

Yet, the pleasures of satisfying personal scientific curiosity aside, it may be difficult to make a compelling case for initiating a massive and costly effort to sequence the human genome by the year 2000 or some other arbitrary, fast-approaching deadline. In the midst of the debate about the wisdom of launching the Human Genome Initiative, this question arises: Would both science and society be better served by a less glamorous, slower-paced, multidisciplinary approach to deciphering the human genome — one that integrated raw DNA sequencing data with the findings of simultaneous studies of human family pedigrees, cell biochemistry and other traditional gene-mapping strategies?

Alternatives

Some scientists have publicly suggested that there would be significant advantages to relying on such a balanced attack as an alternative to HGI. First, the technology of gene sequencing is in rapid flux. There is every reason to assume that new laboratory techniques and computerization of some processes will dramatically reduce the price we will have to pay for each base pair in the human genome. There are already indications that new DNA sequencing technologies on the horizon will be able to perform tens, even hundreds, of times faster than existing ones. By lowering the priority of a totally sequenced genome, science and industry would have additional time to develop cheaper, more efficient, automated devices for carrying out the tedious laboratory work required. Second, by redefining the challenge of human gene mapping as a multidisciplinary, cooperative effort rather than some sort of international contest for economic advantage or scientific prestige, researchers might be encouraged to focus on the most pressing medical problems in human genetics first. Third, buttressing raw sequencing data with a variety of other clues to the complexities of gene activity would render existing data banks of base sequences more meaningful and, at the same time, help guard against any premature applications of sequencing data.

Enthusiasm is what fuels the process of scientific exploration. But — especially in human genetics — enthusiasm must sometimes be tempered by a careful look at research priorities and the possible adverse effects of proceeding too quickly with certain areas of research. The sequencing of the human genome will leave us with a deluge of undigested genetic data. Already the first trickle of this information has almost overwhelmed the first pilot DNA data bases established at research centers in the United States during the early 1980s. And those flows of information will increase manyfold.

Possible Abuses

Because this store of base sequences will represent an incredibly precise inventory of minute, often inconsequential hereditary differences between people, it will be highly vulnerable to abuse. Computerized human gene banks are likely to offer new opportunities for wholesale genetic screening programs — some useful, others of dubious merit — for identifying individuals who harbor genes considered "defective" or "inferior." With such a bounty of unexplored data on the human genome available, there may be a temptation to hastily assign too great a causal role to many freshly mapped genes — simply because of their preliminary statistical associations with perplexing health problems.

As in the past, some people are bound to eagerly exploit such findings by publicly proclaiming that they offer "scientific solutions" for "fixing" everything from alcoholism and mental illness to homosexuality and learning disabilities. In the absence of other, more compelling evidence for links between specific DNA sequences and disease, our swollen DNA data banks could quickly become reservoirs of easy genetic "answers" to complex human problems.

Reductionism
In the end, if we decide to invest limited research resources in a massive project to sequence the human genome by some arbitrary date such as the year 2000, the greatest risk we may face is self-delusion. And this will be especially true if the decision is made at the expense of other, less glamorous, but useful, areas of research in genetics.

History suggests that in the wake of dramatic scientific conquests — such as the testing of the first atomic bomb in Alamogordo, New Mexico; the first U.S. lunar landings; and the building of the first commercial nuclear power plants — we have often congratulated ourselves prematurely on our newfound "mastery" over the forces of nature. In much the same way, the successful completion of a massive, goal-oriented, human DNA sequencing program could give us a false sense of scientific mastery over our species' genome.

In many ways, modern biology has not yet begun to fathom the true complexity of even the simplest living cell. By its very nature, science sets out to explore biological processes by dissecting them into their lifeless ingredients. But by so doing, it unavoidably blinds itself to some of the most subtle interactions between these component parts that routinely take place deep inside all living organisms.

Scientists tend to become most painfully aware of such blind spots when they try to explain the workings of the most complex biological systems — such as the human hereditary apparatus and the human brain, with its tens of billion of interacting nerve cells capable of conjuring up mental experiences ranging from consciousness and artistic creativity to pleasure and pain.

In centuries past, biologists who argued that living things possessed mysterious qualities that would remain forever beyond human comprehension were branded "vitalists." Understandably, most modern biologists, who believe that the underlying processes of all living systems can be reduced to our ordinary notions of chemistry and physics, are not anxious to be associated with the outmoded and discredited views of traditional vitalism.

But a growing number of scientists are beginning to suggest that the most challenging questions in neurobiology and other fields in the life sciences seem to defy conventional scientific answers and that this realization need not condemn them to some mystical, antiscientific worldview. Instead, reputable voices in the scientific community — among them, Nobel Prize-winning brain researcher Roger Sperry — are beginning to propose thought-provoking, alternative views. Is it not possible, they have suggested, that at each layer of complexity in living creatures the physical interaction of what we label constituent parts results in "emergent properties" — qualities that paradoxically seem to exceed the sum of those parts?

Just as nuclear physicists have been forced to resign themselves to the inevitable vagaries of defining subatomic particles, biologists may have to recognize the limitations of their fragmentary visions of genes. In the rapidly unfolding realm of human genetics, this may mean that even a lengthy, hard-won list of DNA bases will provide no more than a glimpse of the fleeting world of living genes.

For example, about a decade ago geneticists succeeded in sequencing the total nucleotide base sequence of the ØX174 virus — all 5,400 bases and nine genes. Recently the entire DNA sequence of the bacterium *E. coli* has been documented — a genetic text totaling about three million bases. Those achievements were viewed, justifiably as a major technical tour de force by most members of the scientific community. Yet, while that information has shed considerable light on the field of microbial genetics, most would agree that we are still a long way from comprehending the full mystery of living viruses and bacteria.

The complete DNA sequence of the human genome is intrinsically far more interesting to us than that of any mere microbe. But we would be deluding ourselves if we thought that by possessing it we could grasp the full meaning of human inheritance — indeed, of our very humanness. The lives of genes are dazzlingly complex dances involving the simultaneous interactions of countless genes, enzymes, metabolic processes and environmental factors. Thus, it would be simplistic to claim that a gene's multidimensional dance could be fully choreographed by a one-dimensional base-pair script — however vital that text is to the outcome of the performance.

Hubris

The zeal with which many scientists seem to desire a completely decoded human genome is testimony to their deep-seated drive to answer some of the most profound questions that still surround the biology of the human organism. And that is always a worthy mission. At the same time, it may also reflect an excessive eagerness to pose questions that we know have relatively easy answers — rather than questions likely to lead to even more difficult questions. That dilemma has been succinctly described by one of genetics' most eloquent spokesmen, French Nobel Prize-winning molecular biologist François Jacob:

For the only logic that biologists really master is one-dimensional. As soon as a second dimension is added, not to mention a third one, biologists are no longer at ease. . . . However, during the development of the embryo, the world is no longer merely linear. The one-dimensional sequence of bases in the genes determines in some way the production of two-dimensional tissues and organs that give the organism its shape, its properties, and . . . its four-dimensional behavior. How this occurs is a mystery.

If we choose to ignore the still-staggering limitations to our knowledge of human inheritance — preferring instead convenient illusion — we could easily fall prey to a most dangerous human folly. DNA sequences in hand, we could find ourselves sliding down a slippery slope toward that perilous state of mind that is a mix of bloated human pride and self-confidence that ancient Greek thinkers referred to as hubris. There is wisdom on this matter in the words of Isaac Bashevis Singer — another Nobel laureate, but a man of literature rather than science:

Our knowledge is a little island in a great ocean of nonknowledge.

EPILOGUE

SEARCHING FOR A NEW MYTHOLOGY

As regards ethics, what is important is to realize the new dangers and to consider what ethical outlook will do most to diminish them.
— Bertrand Russell, in *The Science to Save Us from Science*

A myth has moral content. It carries its own meaning. It secretes its own values. . . . science is an attempt to free investigation and knowledge from human emotional attitudes . . . . This so-called "objective world" thus becomes devoid of mind and soul, of joy and sadness, of desire and hope.
— François Jacob, in *The Possible and the Actual*

The serious problems in life, however, are never fully solved. If ever they should appear to be so it is a sure sign that something has been lost.
— Carl Jung, in *The Structure and Dynamics of the Psyche*

The Limits of Science
Modern genetics has granted us significant new powers over the processes of biological inheritance. It has provided us with techniques for splicing together genes from distantly related species, deciphering the cryptic chemical messages of genetic molecules and analyzing and even altering individual genotypes.

We will continue to reap enormous benefits from this scientific knowledge — in medicine, agriculture, industry and other areas of application. But we will also continue to pay a heavy price for this and other forms of scientific knowledge. For the insights of modern genetics — like those of astronomy, atomic physics and other disciplines — have forced us to reexamine the place of our species in the natural world and even to reassess what it means to be human.

The impact of science and technology on modern society has been so profound that it has shaken the foundations of our myths, religions and other traditional reservoirs of wisdom and moral values for our species. Ironically, the muting of traditional voices of moral authority by modern science has taken place at precisely the same time modern science is forcing us to face a growing number of difficult ethical issues — many of them unprecedented in human history.

Given the extraordinary accomplishments of twentieth-century science, it is not surprising that some of us tend to respond to such issues by substituting a blind faith in scientific progress and in scientific ways of knowing for individual moral responsibility. But that faith is misplaced unless we recognize that science offers a very limited view of the world.

For the most part, it is the business of science to systematically dissect nature into manageable bits and pieces. Most scientists explore the natural world by focusing on one small part of it, isolating that part from the complexities that surround it and studying it under carefully controlled conditions. As history demonstrates, this is an extraordinarily powerful approach to learning about nature, and it has unveiled useful secrets that have already paid rich dividends. But the methods of science do have built-in limitations, and we ignore them at our peril.

In the first place, the fruits of scientific inquiry are not necessarily facts. They are best described as tentative truths — unfinished ideas about nature's workings that are subjected to continuous criticism, modification and rejection by other researchers. In the second place, scientific explanations of the natural world are necessarily fragmentary. For science is condemned to see nature as a mosaic of arbitrarily defined, component parts. The methods of science are simply incapable of encompassing the dazzling complexity of whole systems — whether they are living cells, thinking brains or self-sustaining ecosystems in the tropical forest.

By sacrificing the whole for the part, scientists are able to understand the workings of isolated areas of natural systems and to gain control over some of those processes. When science creates new technological tools, they often yield instant social and financial rewards. But these successes can blind us to the fact that science seldom has much to say about the possible future effects of these applications on our societies and ecosystems. We need look no farther than the long-term effects of the automobile, nuclear fission, oral contraceptives, agricultural pesticides or television to appreciate the myopia of science. The effects of each of these applications of science extend far beyond the immediate uses for which they were intended.

In many ways, the accelerating pace of scientific and technological advances has thrown us out of balance with the natural world. Technology has equipped our species with the mechanical equivalent of muscle power that now greatly exceeds that of any species that has ever lived. It has provided us with the resource base with which to multiply into the largest and most far-flung population of large mammals on earth. And a whole host of human activities — from commercial farming and forestry to the combustion of oil and coal — have altered the face of the planet in ways that are easily visible from outer space.

The tragedy is that in the process most of us have lost any clear sense of our species' place in global ecosystems and of our biological kinship with other living things. We must not lose sight of this larger context as we continue to tinker with genes and shape the hereditary futures of species. For as we embark on this new era of applied molecular genetics, we are in some ways incredibly shortsighted. We are so intent on rushing to exploit our newly acquired insights that we often do not have the faintest idea of the long-term consequences of our technologies.

Science and technology are in themselves neither good nor evil.

They are the product of a human curiosity — the mind's relentless urge to explore, to know, to change. And that is a quality that we must always nourish. But we must also recognize the need for a moral framework within which scientific inquisitiveness can be expressed without exposing human populations and their surroundings to unacceptable risks or irreparable harm.

Such a system would have to be just and broadly acceptable. It would have to be resilient enough to make room for new scientific truths as they emerge. And it would have to embody humanity's most cherished — and most vulnerable — values. Even with such a moral framework in place, we would be naive to expect it to provide answers to every difficult ethical dilemma arising from science. But such a framework might at least inspire us to ask ourselves different kinds of questions of science, to search for different kinds of scientific applications and to seek different kinds of long-term goals than do some of the most influential institutions of moral authority in Western society today.

Genethic Principles — A Review

It would be presumptuous to pretend that one book could begin to point out a comprehensive set of moral guidelines for genetic responsibility — "genethic" principles, as we have called them — that would meet all of the ethical challenges that lie ahead. The clash between genetics technologies and human values has already resulted in a deluge of difficult ethical problems. And it will continue to do so in the future, in ways that are simply impossible to anticipate.

By proposing our own genethic principles for a selection of existing ethical issues generated by modern genetics — from the historic XYY controversy to recent proposals to sequence the human genome — we do not pretend to have resolved them. Nor do we wish to imply that the values contained in our brief inventory of genethic imperatives begin to address all of the difficult moral issues being raised by genetics. But we hope we have demonstrated a process by which nonscientists — armed with an understanding of basic principles of heredity and a sense of some of the uncomfortable lessons of history — can join in a dialogue about critical genethic issues that is too often reserved for individuals with special privileges, powers or expertise.

We have also tried to show how one can, by exploring even a limited number of ethical issues in detail, begin to derive lasting lessons about heredity and the human experience, whose meaning often extends far beyond the specific moral problem under investigation. In this sense, a carefully selected and well-documented case study of friction between genetics and human values can sometimes be elevated to the status of "scientific parable" — at least a provisional one, until history provides us with a clearer, more meaningful one.

Unlike their Old Testament equivalents, the 10 moral principles we propose — nine in previous chapters, one to be presented in this chapter — cannot claim to be either divinely inspired or cast in tablets of stone. They simply represent our best efforts — buttressed by the ideas of many other scientists and bioethicists — to come up with working solutions to some enormously challenging moral questions. The purpose of these principles is not to serve as any final word on the subject but rather to stir a wider public interest in these and other genethic issues that might eventually lead to a genuine consensus.

Here, in review, are the nine genethic principles we have presented so far:

1. *To grasp many of the difficult ethical issues arising from modern genetics, one must first understand the nature of genes — their origins, their role in the hereditary processes of cells and the possibilities for controlling them.*
2. *The vast majority of human hereditary differences are polygenic, or involve the interplay of many genes; therefore, it is a dangerous simplification to proclaim a causal relationship between human behaviors and so-called "defects" in human DNA.*

3. *Information about an individual's genetic constitution ought to be used to inform his or her personal decisions rather than to impose them.*

4. *While genetic manipulation of human somatic cells may lie in the realm of personal choice, tinkering with human germ cells does not. Germ-cell therapy, without the consent of all members of society, ought to be explicitly forbidden .*

5. *The development of biological weapons is a misapplication of genetics that is morally unacceptable — as is the air of secrecy that often surrounds it.*

6. *The information contained in genetic molecules is vulnerable to loss through mutations caused by sunlight, radioactivity, chemicals and other external mutagenic forces. Each of us has a responsibility to develop an awareness of potential mutagens in our immediate surroundings and to seek to minimize environmentally induced damage to our DNA.*

7. *Until we have a better understanding of the extent of genetic exchange between distantly related species in nature, we ought to consider evolutionary "boundaries" — areas of relatively limited genetic exchange — as at least provisional warning signs of potential danger zones for the casual transfer of recombinant genes between species.*

8. *Genetic diversity, in both human and nonhuman species, is a precious planetary resource, and it is in our best interests to monitor and preserve that diversity.*

9. *The accumulation of genetic knowledge alone — however precious in its own right — does not guarantee wisdom in our decisions regarding human heredity; if such knowledge breeds a false sense of human mastery over genes, it can even lead to folly.*

Again, these genethic principles can only be regarded as the barest of beginnings. In the years ahead, we will require a rich assortment of humane, scientifically sound, broadly acceptable genethic principles to help us resolve all sorts of other difficult ethical questions concerning the directions we are willing to allow new genetic technologies to take. We acknowledge that at this very moment there are a number of pressing ethical issues that cry out for our attention. We have simply chosen not to address these questions in detail here because of limitations of time and space. Their absence does not in any way diminish their importance.

For example, to what extent should individuals and private corporations be permitted to use patents to claim exclusive "ownership" of information contained in the genomes of genetically engineered organisms? Should profit-seeking biotechnology corporations be permitted to pursue genetic applications shaped almost entirely by profit motives — at the expense of less financially rewarding but more compassionate concerns such as agriculture, health and nutrition in developing nations? Should we invest huge sums of money in the research and development of expensive techniques to manipulate genes associated with extremely rare hereditary illnesses — when millions of children in Third World countries are suffering from diseases that could easily be prevented, using scientific knowledge that already exists, simply by altering specific environmental conditions? In many cases, for instance, permanent blindness resulting from dietary vitamin A deficiency could be prevented by providing children with inexpensive vitamin supplements. And, in a more philosophical vein, what psychological price are we likely to pay if we decide to use new genetic technologies not to celebrate the genetic diversity we uncover in human beings but to impose arbitrary "definitions" of what it means to be human? These and countless other genethic issues also need to be communicated to the public with the same intensity with which we have explored a select few.

The Need for a New Mythology
Where can we turn for moral guidance as we try to cope with this avalanche of difficult and disturbing moral issues? There is no easy answer to this question. As each of us grapples with the ethical issues arising from modern genetics, we will rely, in part,

on our own consciences. We can also draw on the wisdom and humanity embodied in our traditional religions, political parties, professional organizations and courts of law. And we can seek the thoughtful counsel of trusted individuals — scientists, physicians, philosophers, religious and political leaders and others.

But should we seek guidance only from the traditional moral authorities of the Western societies that gave birth to modern genetics? We believe that is an unrealistic and ethnocentric view. For that reason, we offer our tenth and final genethic principle. In a sense, it can be seen as a moral imperative that contains within it a formula for creating future genethic principles as they are required:

10. *The search for meaningful ethical principles to help guide us through difficult personal and collective decisions arising from the applications of modern genetics will be an endless process. To succeed, it must lead us beyond the rigid boundaries of Western science and even Western philosophical thought to rich, cross-cultural realms that embrace other ways of knowing.*

By "other ways of knowing," we are referring to sources of human wisdom, lying beyond the mainstream of Western cultural and scientific tradition, that have something to contribute to important genethic issues. To tap these non-Western moral resources, we will have to rely increasingly on a new breed of scholar — cultural anthropologists, schooled not just in the intricacies of non-Western religion, mythology, literature and folklore, but also in the latest advances in molecular genetics. Their task will be to sift patiently through mounds of living, cross-cultural knowledge for fragments of truth that might help us along our way. That search might encompass, for example, the ancient writings of Buddhist or Hindu scholars, the sacred texts of Islam or the worldviews of American Indian peoples like the Quechua of ancient Peru, the Maya of central America or the Haida of the Pacific Northwest. To a limited extent, this process has already begun. But except among small circles of philosophers and bioethicists, the notion of seeking solutions to difficult genethic issues not just in our own cultural backyard but also in the accumulated wisdom of all humankind has not yet received broad acceptance.

The aim of such efforts would be to forge a new, cross-cultural synthesis of moral values that addresses the most disturbing questions raised by modern molecular genetics and genetic engineering. It would be forged from elements drawn from the rich diversity of myths, rituals and other expressions of human values concerning such timeless topics as heredity, life and death, and the natural world. Out of a mingling of the moral vision of many cultures and the latest insights of molecular genetics might gradually emerge what could be called a new and scientifically relevant mythology for our times — one steeped not only in the myth-shattering truths of science but also in the values that might help us use that scientific knowledge wisely and humanely.

What form might the creative, cross-cultural synthesis of science, religion and other ways of knowing take? Because twentieth-century science has dropped us so suddenly into the unexplored ethical territory of molecular genetics and genetic engineering, it is impossible to know what sort of moral maps we will devise to find our way.

But scholars who have devoted their lives to the comparative study of cultural symbols and myths have offered tantalizing clues. They have found compelling evidence that myths — in one form or another — encode a number of universal themes. The messages of myths have served society by nourishing creativity, cohesiveness and moral order in human communities throughout history and in a diversity of geographic settings. Knowing this, we might reasonably expect that new generations of myths embodying the genethic values we so desperately require may surface through the same ponderous birth throes of cultural evolution that have always given rise to and reformulated myths.

Precisely what messages and mores our new science-sensitive myths may harbor remain a mystery for now. But there is some pleasure in anticipating what their combined effect on us is likely to be. For to be relevant, this new mythology will have to evoke in human beings — scientists and nonscientists alike — an abiding sense of awe and humility toward all biological systems, whose stunning complexities continue to outstrip our richest imagination.

GLOSSARY

allele   One of two or more alternative forms of a gene that exist at a specific gene location on a chromosome, giving rise to alternative hereditary characteristics.

amino acid   The base chemical subunit of proteins; there are 20 common amino acids.

amniocentesis   The removal and analysis of a small quantity of amniotic fluid containing fetal cells during pregnancy.

anaphase   An intermediate stage in cell division during which chromosomes move to opposite poles of the spindle apparatus.

antibody   A protein produced by the immune system that binds specifically to a particular foreign substance, or antigen.

anticodon   A nucleotide base triplet in a transfer RNA molecule that pairs with a complementary base triplet, or codon, in a messenger RNA molecule.

antigen   A substance that stimulates the immune system to produce specific antibodies against it.

autosome   Any chromosome that is not a sex chromosome.

bacteriophage   A virus that infects bacteria cells.

centimorgan   A unit of distance between genes on a chromosome based on the frequency of crossing over between those genes during meiosis.

central dogma The concept that in nature hereditary information generally flows unidirectionally from DNA molecules to RNA molecules to proteins.

centriole A specialized cell organelle that plays a crucial role in the organization of spindle fibers during cell division.

centromere Chromosomal region to which chromatids and spindle fibers attach during cell division.

chemical evolution The gradual development of complex chemical compounds from simpler chemicals by a process analogous to natural selection.

chorionic villi sampling The removal and analysis of fetal cells from chorionic membranes during pregnancy.

chromatid One of the two parallel threads of DNA, joined together by a centromere, that make up a replicated chromosome.

chromatin The diffuse substance of chromosomes.

chromophore The naked nucleic acid molecule that is the primary gene-bearing structure of prokaryotic cells.

chromosome The condensed rod made up of a linear thread of DNA interwoven with protein that is the gene-bearing structure of eukaryotic cells.

cistron A DNA sequence that codes for a specific polypeptide; a gene.

codominance A condition in which a heterozygote simultaneously shows the phenotypic effects of both alleles.

codon A nucleotide base triplet of DNA that codes for an amino acid.

conjugation The joining of two bacterial cells, during which genetic material is transferred from a donor to a recipient cell.

crossing over The exchange of corresponding chromosomal segments during cell division.

deoxyribonucleic acid   See **DNA.**

differential reproduction   The unequal reproductive success rates of organisms with different genotypes.

diploid cell   A cell containing two full sets of chromosomes.

DNA (deoxyribonucleic acid)   The gene-bearing double helix molecule, made up of linked nucleotide subunits, that is the primary hereditary molecule in most species.

DNA ligase   An enzyme that can rejoin nucleotides in a DNA strand.

DNA polymerase   An enzyme that can synthesize a new DNA strand using an existing DNA strand as a template.

DNA polymorphism   The occurrence, in a population, of multiple alternative DNA sequences at a particular gene site.

DNA sequencing   The process of deciphering the precise order of nucleotide bases in a DNA molecule.

dominant   Said of an allele that expresses its phenotype even in the presence of a recessive allele.

electrophoresis   The separation and identification of molecules based on their movement through an electrically charged field.

eugenics   A strategy of trying to orchestrate human evolution through programs aimed at encouraging the transmission of "desirable" traits and discouraging the transmission of "undesirable" ones.

eukaryote   An organism with cells characterized by a true nucleus, specialized cell organelles and mitotic cell division.

exon   A DNA sequence that is ultimately translated into protein.

express   To transcribe or translate a gene's message into a molecular product.

gamete   A haploid reproductive cell — e.g., a sperm or an egg cell in mammals — in sexually reproducing organisms.

gene The basic physical and functional unit of heredity that is transmitted from one generation to the next and can be transcribed into a polypeptide or protein.

gene cloning The synthesis of multiple copies of a particular DNA sequence using a bacteria cell or another organism as a host.

gene frequency The percentage of alleles of a given type in a population of organisms.

gene insertion The addition of one or more copies of a normal gene into a defective chromosome.

gene linkage The hereditary association of genes located on the same chromosome.

gene modification The chemical repair of a gene's defective DNA sequence.

gene scanning The process of systematically scanning individual genotypes for possible hereditary defects or abnormalities.

gene surgery The precise removal of a defective DNA sequence and its replacement by a normal sequence.

gene therapy The medical replacement or repair of defective genes in living cells.

genetic code The set of relationships between the nucleotide base-pair triplets of a messenger RNA molecule and the 20 amino acids that are the building blocks of proteins.

genetic drift Random variation in gene frequency from one generation to the next.

genetic engineering The technique of altering the genetic makeup of cells or individual organisms by deliberately inserting, removing or altering individual genes.

genetic monitoring tests Tests that examine hereditary molecules for early indications of genetic damage or disease.

genetic screening The process of systematically scanning individual genotypes for possible hereditary defects or abnormalities.

genome The entire genetic endowment of an organism or individual.

genotype An individual's genetic makeup underlying a specific trait or constellation of traits.

germ cell Reproductive cell.

globins The protein subunits of biologically important molecules such as hemoglobin.

haploid cell A cell containing only one set, or half the usual diploid number, of chromosomes.

heme group The crucial, oxygen-binding chemical component of hemoglobin molecules found in red blood cells.

heterochromatin Condensed, dark-staining regions of chromosomes thought to be for the most part genetically inactive.

heterosis The tendency for heterozygous hybrid organisms to display greater vigor than their homozygous parents.

heterozygote An individual whose genotype is characterized by two different alleles of a gene.

histones Proteins associated with chromosomal DNA.

homologous chromosomes Matched chromosomes that are virtually identical in shape, size and function and that pair during meiotic cell division.

homozygote An individual whose genotype is characterized by two identical alleles of a gene.

incomplete dominance A condition in which a heterozyote offspring has a phenotype that is distinct from, and intermediate to, homozygous, parental phenotypes.

industrial melanism   The natural selection of darker, or melanic, color forms in certain species as a result of industrial pollution.

insertional mutations   Changes in the base sequence of a DNA molecule resulting from the random integration of DNA from another source.

interphase   A resting stage of cell division preceding prophase, in which DNA exists in a diffuse state rather than as condensed chromosomes.

intron   A noncoding DNA sequence within a gene that is initially transcribed into messenger RNA but later snipped out.

ionizing radiation   Radiant energy capable of breaking down molecules in its path into charged particles, or ions.

isotope   One of the possible alternative forms of a chemical element sharing the same number of protons but having a different number of neutrons.

karyotype   The full array of chromosomes in a cell or an individual, visible through a microscope during mitotic metaphase.

kilobase   A segment of DNA 1,000 bases long.

linkage   The tendency of genes located in close proximity to one another on a chromosome to be inherited together.

liposomes   Membrane-bound vesicles constructed in the laboratory to transport biological molecules.

meiosis   A mode of cell division in eukaryotes in which a diploid parent cell gives rise to haploid reproductive cells, or gametes.

messenger RNA   An RNA molecule that has been transcribed from a gene-bearing DNA molecule and will later be translated into a protein.

metaphase   A stage in cell division in which chromosomes align along the metaphase plate of the dividing cell.

mitosis A mode of cell division in eukaryotes in which a single parent cell gives rise to two genetically identical daughter cells.

molecular genetics The study of the molecular basis of gene structure and function.

monoculture The agricultural practice of cultivating crops consisting of genetically similar organisms.

monogenic Controlled by or associated with a single gene.

mutagen An agent that increases the mutation rate.

mutation A heritable change in a DNA molecule.

natural selection An evolutionary process by which those organisms that are best adapted to a particular environment tend to leave a greater number of offspring.

nondisjunction The failure of homologous chromosomes to separate during meiosis, resulting in daughter cells with missing or surplus chromosomes.

nucleic acid One of a number of acidic, linear, information-bearing, biological molecules belonging to one of two chemical families — DNA or RNA.

nuclein An early term for the unknown nucleic acid content of cell nuclei.

nucleotide A chemical subunit composed of a sugar, phosphate and base, which makes up the nucleic acids DNA and RNA.

oncogene A gene that causes cancer.

oocyte A diploid cell destined to undergo meiosis to produce a haploid egg.

oogenesis The process by which egg cells arise from diploid sex cells.

oogonium A diploid cell destined to undergo mitosis to form an oocyte.

opine A chemical compound, synthesized by tumorous plant cells infected with crown gall bacteria, that serves as a nutrient for the bacteria.

organelle A cell structure, such as the mitochondrion and the chloroplast, that carries out a specialized function in the life of a cell.

palindrome A DNA sequence that is the same when one strand is read left to right and the other is read right to left.

phage See **bacteriophage.**

phenotype The detectable characteristics associated with a particular genotype.

plasmid A circular, self-replicating form of DNA found in many species of bacteria that can sometimes be used as a vector to shuttle recombinant genes into another species.

point mutation A mutation that results in the substitution of one nucleotide base for another.

polar body A relatively tiny daughter cell arising from the first or second meiotic division in females.

polygenic Controlled by or associated with more than one gene.

polymer A chemical substance formed by the chainlike linkage of simpler molecules.

polymorphisms Variant forms of a particular gene that occur simultaneously in a population.

polynucleotide A linear chain of nucleotide molecules.

polypeptide A protein or portion of a protein consisting of two or more amino acid molecules.

population A local group of organisms belonging to the same species and capable of interbreeding.

population genetics The study of inheritance at the level of populations.

transcription   The synthesis of messenger RNA from a DNA template.

transfer DNA   See **T-DNA.**

transfer RNA   A class of small, mobile RNA molecules that transport specific amino acids to the ribosome during protein synthesis.

transgenic   Said of an organism produced by the transfer and expression of genes from another species.

translation   The synthesis of protein from a messenger RNA template.

transposon   A highly mobile DNA sequence that moves from one chromosomal location to another.

trisomy   A condition in which cells in an individual possess a full diploid set of chromosomes, plus a surplus chromosome — as in trisomy 21, or Down's syndrome.

variation   Differences in the frequency of genes and traits among individual organisms within a population.

vector   The plasmid, virus or other vehicle used to shuttle a cloned DNA sequence into the cell of another species.

zygote   A fertilized egg.

FOR FURTHER READING

Chapter 1

Chai, Chen Kang. *Genetic Evolution*. Chicago: University of Chicago Press, 1976.

Dillon, Lawrence S. *The Genetic Mechanism and the Origin of Life*. New York: Plenum Press, 1978.

Dunn, L. C. *A Short History of Genetics*. New York: McGraw-Hill, 1965.

Hoagland, Mahlon. *Discovery: The Search for DNA's Secrets*. New York: Van Nostrand Reinhold, 1981.

Margulis, Lynn. *Early Life*. Boston: Science Books International, 1982.

Sturtevant, A. H. *A History of Genetics*. New York: Harper and Row, 1965.

Tiley, N.A. *Discovering DNA: Meditations on Genetics and a History of the Science*. New York: Van Nostrand Reinhold, 1983.

Vogel, Friedrich, and Arno G. Motulsky. *Human Genetics: Problems and Approaches*. Heidelberg: Springer-Verlag, 1979.

Chapter 2

De Duve, Christian. *A Guided Tour of the Living Cell*. New York: W. H. Freeman and Company, 1984.

Maxson, Linda R., and Charles H. Daugherty. *Genetics: A Human Perspective*. Dubuque, Iowa: Wm. C. Brown, 1985.

Readings from "Scientific American" Genetics. San Francisco: W. H. Freeman and Company, 1981.

Chapter 3

Clark, Brian F. C. *The Genetic Code and Protein Biosynthesis.* Institute of Biology's Studies in Biology series, No. 83. 1979. London: Edward Arnold, 1979.

De Duve, Christian. *A Guided Tour of the Living Cell.* New York: W. H. Freeman and Company, 1984.

Klug, William S., and Michael R. Cummins. *Concepts of Genetics.* Columbus, Ohio: Merrill, 1986.

Maxson, Linda R., and Charles H. Daugherty, *Genetics: A Human Perspective.* Dubuque, Iowa: Wm. C. Brown, 1985.

Readings from "Scientific American" Genetics. New York: W. H. Freeman and Company, 1981.

Chapter 4

Bishop, J. A., and L. M. Cook, eds. *Genetic Consequences of Man Made Change.* London: Academic Press, 1981.

Chai, Chen Kang. *Genetic Evolution.* Chicago: University of Chicago Press, 1976.

Dodson, Edward O., and Peter Dodson. *Evolution: Process and Product.* Boston: Prindle, Weber and Schmidt, 1985.

Dobzhansky, Theodosius. *Mankind Evolving.* New Haven, Conn.: Yale University Press, 1962.

Lewontin, Richard C. *Human Diversity.* New York: Scientific American Books, 1982.

Chapter 5

Office of Technology Assessment, U.S. Congress. *Impacts of Applied Genetics: Micro-organisms, Plants, and Animals.* Washington, D.C.: U.S. Government Printing Office, 1981.

Old, R. W., and S. B. Primrose. *Principles of Gene Manipulation.* Berkeley: University of California Press, 1981.

Readings from "Scientific American": Recombinant DNA. San Francisco: W. H. Freeman and Company, 1978.

Watson, James D., Nancy H. Hopkins, Jeffrey W. Roberts, Joan A. Steitz and Alan M. Weiner. *Molecular Biology of the Gene.* 4th. ed. Menlo Park, Calif.: Benjamin/Cummings, 1987.

Watson, James D., John Tooze and David T. Kurtz. *Recombinant DNA: A Short Course.* New York: W. H. Freeman and Company, 1983.

Chapter 6

Beckwith, Jon, and Jonathan King. "The XYY syndrome: a dangerous myth." *New Scientist,* November 1974, 474-76.

Gould, Stephen Jay. *The Mismeasure of Man.* New York and London: Norton, 1981.

Kevles, Daniel K. *In the Name of Eugenics: Genetics and the Uses of Human Heredity.* New York: Knopf, 1985.

Lappé, Marc. *Genetic Politics.* New York: Simon and Schuster, 1979.

Walzer, Stanley, Park S. Gerald and Saleem A. Shah. "The XYY Genotype." *Annual Review of Medicine* 29 (1978): 563-70.

Chapter 7

Holden, Constance. "Looking at Genes in the Workplace." *Science* 217 (July 23, 1982): 336-37.

Hubbard, Ruth, and Mary Sue Henefin. "Genetic Screening of Prospective Parents and of Workers: Some Scientific and Social Issues." *International Journal of Health Services* 15, no. 2 (1985): 231-51.

Hunt, Morton. "The Total Gene Screen." *The New York Times Magazine,* January 19, 1986, 33-61.

Lappé, Marc. *The Broken Code: The Exploitation of DNA.* San Francisco: Sierra Club Books, 1984.

Lappé, Marc. "The New Technologies of Genetic Screening." *Hastings Center Report* 14, no. 5 (October 1984): 18-21.

Murray, Thomas H. "Genetic Testing at Work: How Should It Be Used?" *Technology Review,* May-June 1985, 51-59.

Office of Technology Assessment, U.S. Congress. *The Role of Genetic Testing in the Prevention of Occupational Disease.* Washington, D.C.: U.S. Government Printing Office, 1983.

Chapter 8

Bunn, H. Franklin, Bernard G. Forget and Helen M. Ranney. *Human Hemoglobins.* Philadelphia: W. B. Saunders, 1977.

Lewis, Ricki. "Beating the Genetic Odds." *High Technology* 4 (12): 73-78.

Motulsky, Arno G. "The Impact of Genetic Manipulation on Society and Medicine." *Science* 219 (January 14, 1983): 135-40.

Office of Technology Assessment, U.S. Congress. *Human Gene Therapy: A Background Paper.* Washington, D.C.: U.S. Government Printing Office, December 1984.

President's Commission for the Study of Ethical Problems in Medicine and Biomedical and Behavioral Research. *Splicing Life: A Report on the Social and Ethical Issues of Genetic Engineering with Human Beings.* Washington, D.C.: U.S. Government Printing Office, November 1982.

Chapter 9

Harris, Robert, and Jeremy Paxman. *A Higher Form of Killing: The Secret Story of Gas and Germ Warfare.* London: Chatto and Windus, 1982.

Hersh, Seymour M. *Chemical and Biological Warfare: America's Hidden Arsenal.* Indianapolis: Bobbs-Merrill, 1968.

Piller, Charles. "DNA — Key to Biological Warfare?" *The Nation,* December 10, 1983, 597-601.

Stockholm International Peace Research Institute. "CB Weapons Today." Vol. II, *The Problem of Chemical and Biological Warfare.* New York: Humanities Press, 1973.

Suzuki, David, Eileen Thalenberg and Peter Knudtson. *David Suzuki Talks about AIDS.* Toronto: General Paperbacks, 1987.

Wright, Susan, and Robert L. Sinsheimer. "Recombinant DNA and Biological Warfare." *Bulletin of Atomic Scientists,* November 1983, 28.

Zochlinski, Howard. "Biotechnology in Military R & D: Two Firms Look for Future Profits in Joint Ventures." *Genetic Enginering News,* March/April 1983, 28-29.

Chapter 10

Ames, Bruce. "Environmental Chemicals Causing Cancer and Mutations." In *Genes, Cells, and Behavior: A View of Biology Fifty Years Later,* edited by Norman H. Horowitz and Edward Hutchings, Jr., 50th Anniversary Symposium, Division of Biology, California Institute of Technology, November 1-3, 1979. San Francisco: W. H. Freeman and Company, 1980.

Hartl, Daniel L. *Human Genetics.* New York: Harper and Row, 1983.

Howard-Flanders, Paul. "Inducible Repair of DNA." *Scientific American* 245, no. 5 (November 1981): 72-80.

Levine, Louis. *Biology of the Gene.* 3d ed. St. Louis: Mosby, 1980.

Stine, Gerald James. *Biosocial Genetics: Human Heredity and Social Issues.* New York: Macmillan, 1977.

Suzuki, David T., Anthony J. F. Griffiths, Jeffrey H. Miller and Richard C. Lewontin. *An Introduction to Genetic Analysis.* 3d ed. New York: W. H. Freeman and Company, 1986.

Chapter 11

Chilton, Mary-Dell. "A Vector for Introducing New Genes into Plants." *Scientific American* 248, no. 6 (1983): 50-59.

Dobzhansky, Theodosius, Francisco J. Ayala, G. Ledyard Stebbins and James W. Valentine. *Evolution.* San Francisco: W. H. Freeman and Company, 1977.

Krimsky, Sheldon. "Breaching Species Barriers." In *Genetic Alchemy: The Social History of the Recombinant DNA Controversy.* Cambridge, Mass.: MIT Press, 1982.

Readings from "Scientific American": Industrial Microbiology and the Advent of Genetic Engineering. San Francisco: W. H. Freeman and Company, 1981.

Sinsheimer, Robert. "An Evolutionary Perspective for Genetic Engineering." *New Scientist* 73 (January 20, 1977): 150-52.

Chapter 12

Asturias, Miguel Angel. *Men of Maize.* Translated by Gerald Martin. New York: Dell, 1975.

Barrett, J. A. "The Evolutionary Consequences of Monoculture." In *Genetic Consequences of Man Made Change,* edited by J. A. Bishop and L. M. Cook. New York: Academic Press, 1981.

Heiser, Charles B. *Seeds to Civilization: The Story of Food.* San Francisco: W. H. Freeman and Company, 1981.

Harlan, Jack R. *Crops and Man.* Madison, Wis.: American Society of Agronomy, 1975.

Mangelsdorf, Paul C. *Corn: Its Origin, Evolution, and Improvement.* Cambridge, Mass.: Harvard University Press, Belknap Press, 1974.

National Research Council. *Conservation of Germplasm Resources: An Imperative.* Washington, D.C.: National Academy of Sciences, 1978.

Baer, Adela S. *The Genetic Perspective.* Toronto: W. B. Saunders, 1977.

Chapter 13

Baskin, Yvonne. "GenBank: Storehouse for Life's Secret Code." *Science Digest,* May 1983, 94-95.

Brennan, James R. *Patterns of Human Heredity: An Introduction to Human Genetics.* Englewood Cliffs, N.J.: Prentice-Hall, 1985.

Lewin, Roger. "Proposal to Sequence the Human Genome Stirs Debate." *Science,* June 1986, 1598-1600.

McKusick, Victor A. "The Anatomy of the Human Genome." *The Journal of Heredity*, 71, no. 6 (November-December 1980): 370-91.

McKusick, Victor A. "The Mapping of Human Chromosomes." *Scientific American* 224, no. 4 (April 1971): 104-14.

Vogel, F., and A. G. Motulsky. *Human Genetics: Problems and Approaches.* 2d ed. Berlin: Springer-Verlag, 1986.

Epilogue

Campbell, Joseph. *Creative Mythology: The Masks of God.* New York: Penguin Books, 1976.

Campbell, Joseph. Myths to Live By. New York: Bantam Books, 1973.

Eliade, Mircea. *From Primitives to Zen: A Thematic Sourcebook on the History of Religions.* San Francisco: Harper and Row, 1977.

Stubbe, Hans. *History of Genetics.* Translated by T. R. W. Waters. Cambridge, Mass.: MIT Press, 1973.

INDEX

abortion
  hereditary disease and, 166, 192, 206
  XYY genotype and, 152, 157-159
ADA deficiency, 186, 193
Afghanistan, 216
Africa, 234
  AIDS virus, as source of, 233
  monoculture in, 291
  sickle-cell anemia in, 186, 187
  sickle-cell trait in, 201
    advantages of, 201-202
aggression
  XYY genotype and, 147-148, 149-150
  Y chromosome and, 149, 159
"Aggressive Behavior, Mental
  Sub-normality and the XYY Male"
  (Jacobs), 147-151
agriculture, 32, 33, 34
  artificial selection in, 34
  biological weapons against, 227
  genetic engineering in, 267, 274-279
    risks of, 280-281
AIDS virus, 232, 233
  African origin of, 233
  biological weapons research and,
    reports on, 233
  groups disproportionately affected
    by, 234
  hypothetical biological warfare
    scenario and, 232-235
albinism, 184
Alzheimer's disease, 330
Ames, Bruce, 258
Ames test, 258-259
amino acids, 29-31, 52-63
  phenylalanine and PKU, 167-168
  protein, building blocks for, 31,
    52-63, 65, 331
  in sickle-cell disease, 189-191
amniocentesis, 166
  sickle-cell anemia and, 192
anthrax
  outbreak of, in Soviet Union, 216
  "promise" of, as biological weapon,
    209, 216
  research on, during World War II,
    214-215
  use of, in World War II, 213

Argentina, 250
Aristotle, 34, 45
artificial epidemic, 231-235
artificial selection, 300
  early genetic engineering, as, 114-115
  early human understanding of, 33
  extrapolation of, to humans, 34
  first systematic studies of, 35
  maize and, 295, 296, 301, 302
    enslavement of, 296, 313
  Mendel's work on, 35-38
Asia, 291
Asilomar Conference, 44, 227-229
  biological weapons on agenda,
    absence of, 227
  Plasmid Working Group and, 228
Asturias, Miguel Angel, 290
Australia, 152, 224
  soldiers of, as subjects of biological
    weapons research, 215
Avery, Oswald, 42
Aztec civilization
  maize and, 293, 294, 296

Bacon, Francis, on knowledge, 180
Beckwith, Jonathan, and XYY controversy, 154-155
Bible, 34, 344
biological weapons (warfare), 208-237
  biohazards of, 227-235
  disadvantages of, 218-219
  disarmament of, 229
  efforts to reduce, 231
  examples of, 209-211
  fear as motive for developing, 220
  future of, 221-227
    agricultural weapons, 227
    biological toxins, mass production
      of, 224-225
  early-warning systems, 226-227
  ethnic weapons, 223-224
  genetically engineered vaccines, 225
  hypothetical scenario of, 231-235
  genethic principle on, 208, 346
  history of, 212-216
  military advantages of, 217-218
  moral unacceptability of, 208, 218,
    234, 345

need to eliminate, 236-237, 345
"offensive" vs. "defensive," 220-221
research on, U.S. funding for,
    216-217
secrecy surrounding, 208, 212,
    214-216, 221, 232, 236-237, 345
treaties forbidding, 213, 228-231
Biological Weapons Convention of
1972, 229-231
criticisms of, 230
deterrent value of, 236
Geneva Protocol, 213
Blake, William, 70

Borges, Jorge Luis, on Dante, 141
boundaries, evolutionary, 265, 269, 273,
    274, 278, 281, 282, 284, 286
in bacteria, molecular, 283-284
Boyer, Herbert, 116
Brinster, Ralph, and gene transfer,
    182, 202

caffeine, as possible mutagen, 240
Cambodia, 216
Canada
    biological weapons and, 214
    genetic mutations and, 263
    genetic screening in, for XYY
        genotype, 153
    maize and, 294
cancer
    chemical mutagens and, 257, 260-261
    delayed onset of, caused by mutation,
        248, 254, 255, 260
    early diagnosis of, potential for, 330
    genes associated with, 318
    hereditary, monogenic diseases as
        cause of, 249
    leukemia, and bone marrow
        transplants, 194, 200
    mutational damage as cause of, 249,
        262
    nuclear explosion and, 255
    in plants, 266. See also crown gall
        bacterium
    Ti plasmids as cause of, 273, 277
    radioactive wastes and, 251
    retinal, visible on chromosome, 322
    skin, and UV damage, 246, 250, 261
    tobacco and, 257, 261, 264
    transformation into, of healthy cell,
        198-199
    treatment of, by cloning of toxins,
        217
Carstairs hospital, and XYY
    controversy, 145, 148

cauliflower mosaic virus, 279
cell division, 38, 42, 67, 68-69, 70-93.
    See also meiosis; mitosis
Central America
    maize cultivation in, 293, 294
central dogma, 53-55
Chernobyl, 254
China
    ancient, and biological warfare, 212
    Japanese biological weapons research
        in, 215
chromosomes
    abnormalities in, radiation-induced,
        255-256
    disease-causing defects in, 144-145
        prenatal inspection for, 166
    dissection of, chemical, 118, 123, 166
    DNA replication on, 67-69
    early understanding of, 38, 39, 45
    karyotyping of, 75, 143-145, 320
        staining for, 75, 144, 317, 320
    meiosis in, 80-93, 142. See also
        meiosis
        crossing over, 86-87, 317, 319
        mutational damage and, 244
    mitosis in, 71-80. See also mitosis
    mutations in, from distributional
        errors, 248, 249
    nondisjunction of, in XYY genotype,
        142, 169
    sex. See X and Y chromosomes
    structure of, 48-50, 74-76
Churchill, Winston, 214
cloning. See Genes, recombinant
Clowes, Royston C., 228
Code of Ethics in Wartime, 208
Cohen, Stanley, 116, 228
Columbus, Christopher, 299
compulsory sterilization
    American eugenics movement and,
        40, 45
    of mental patients, 206
Crick, Francis, 42, 50, 116
crown gall bacterium (Agrobacterium
    tumefaciens) 265-269, 271-289
    action of, 267-268, 273-274
    agricultural genetic engineering and,
        267
    discovery of, 266
    as exception to natural law, 284-287
    as metaphor, 287
    potential of, in recombinant
        technology, 275, 278, 279
    risks of using, 281
    technical problems of, 276-279
Cuba, 299
Curtiss, Roy, III, 228

cystic fibrosis, 184, 330
Darwin, Charles, 35, 39, 45, 95-96
  Origin of Species and, 95
  theory of, foundation of, 95
Darwinian theory, 95-96
Denmark, 153
DNA
  artificial synthesis of, 44, 136-138
    interference in, for gene
      sequencing, 134-135
    interference in, for gene synthesis,
      136-137
  compared with RNA, 53
  defined, 26
  discovery of, 42, 45
  double-helix model of, 42-44, 45, 50,
    51, 116, 317
    as basis for self-replication in, 42,
      50-53, 65-69, 72
  dynamics of
    replication, 65-69, 90, 112, 134,
      242, 244, 246
      interference with, for sequencing,
        134
      mutational damage and, 244, 247
    transcription, 55-58, 65, 70, 78,
      121, 125, 196, 242, 270, 277, 331
    translation, 58-62, 65, 70, 197, 277
  environmental damage to, 169,
    238-264. *See also* mutagens;
    mutation; radiation
    chemical, 257, 260-261
    human insensitivity to, 261-262, 264
    ionizing radiation and, 251-252, 256
    solar (UV) radiation and, 242-244
      repair systems for, 242-248, 259
  genetic screening and, 164, 169, 170.
    See also genetic screening
  manipulation of, for gene therapy,
    183-185. *See also* gene therapy
    in gene insertion, 184, 185
    modification of, *in situ*, 185
    transplantation of, 185
  manufacture of, by cells, 42
  origin of, 29
  prokaryotic, 270
  recombinant, technology of, 44,
    114-140
    biological weapons and, 211-212,
      216-217, 219, 220-235. *See also*
      biological weapons
      approval of research on, in U.S.,
        217
      coverage of, under Biological
        Weapons Convention of 1972,
        230
      hypothetical scenario of, 232-235
    cloning of genes with, 116-132,
      182, 318, 325. *See also* genes,
      recombinant technology of
    diagnostic, for sickle-cell anemia,
      192
    for mapping human genome,
      325-327, 328-339
    mutation identification and, 249
    potential of, 138-140, 182
    regulatory institutions for, 217
  recombination of, natural, 87-89, 92,
    104, 106. *See also* meiosis
  role of, in protein synthesis, 32,
    52-54, 55-64
  sequencing of, 44, 132-137, 335
    in human genome, 139, 165, 318,
      325, 326-327, 328-339. *See also*
      human genome
    insights from, 138-140
    radioactive labelling and, 134-137,
      325
  as source of genetic information,
    53-54
  structure of, 42, 48-55
    constituent nucleotide bases in,
      42-45, 50-55
    chromosome staining and, 322
    in eukaryotic DNA, 269-270
    gene synthesis and, 136-138
    human genome and, 317
    mutation and, 104, 242, 247, 326
    natural recombination and, 87
    restriction enzymes and, 119-122
    sequencing of, 132-136, 325, 338
  transfer (T-) DNA, 268, 273, 275,
    277, 288
Dobzhansky, Theodosius, on evolution,
  94
dominance
  codominance, 111
  incomplete, 110-111
  Mendel's discovery of, 36
  mutations and, 198
  peppered moths and, 98-101
Down's syndrome
  chromosomal mutation, as example
    of, 248
  compared with XYY genotype, 148
  extra autosome in victims of, 144-145
  genetic screening and, 164
Durrell, Lawrence, 160

Eliot, George, 48
emphysema, and genetic screening, 172
environmental toxins. *See also*
mutagens, environmental.
  agricultural genetic engineering and,
    280

of maize, 310
biological weapons as source of, 219
industrialization as source of, 241,
  250, 256-258, 262, 264
radioactive, 252-256
workplace, 161, 174-176
  genetic screening and, 174-175, 176,
    179-180
Escherichia coli
DNA sequence of, 338
gene cloning experiments with,
  127-128
in biological warfare, 225
restriction enzyme in, 118, 122
eugenics, 38-40, 44-45
in Europe, 40
Galton on, 39
hypothetical biological warfare
  scenario and, 234
lessons of, for gene therapy, 204
in Nazi Germany, 41, 44-45
in North America, 40-41, 45
  compulsory sterilization and, 40, 45
  U.S. Immigration Act of 1924 and,
    41, 45
XYY genotype and, 150, 152
eukaryotes, 265, 269, 274, 277, 278,
  282, 284, 286, 287, 331
compared with prokaryotes, 269-270
similarity of, to crown gall bacteria
  T-DNA, 288
Europe, 233
industrial melanism in, 101
monoculture and, 291
northern, and Chernobyl accident,
  254
evolution. *See also* heredity
biological, 27, 30-31, 45, 46-47, 102
  boundaries established by, 265,
    284, 334, 345
chemical, 29
Cohen on, 114
comparative, 318
"continuity of divergence" in, 286,
  287
cultural, and myths, 348
Darwin's theory of, 96
Dobzhansky on, 94
essence of, 95, 102
eukaryote, 271
future, inability of, to anticipate, 174,
  262
genetic drift and, 106, 109
genetic migration and, 107, 109
genetic variability as basis of, 30,
  82-83, 87, 96, 103-106, 304
Hardy-Weinberg equilibrium and,

107-109, 110
human, 111, 301, 330, 333
  environmental toxins and, 175
  genetic damage and, 261
  insights into, through mapping
    human genome, 332-334
irreversibility of, 112
of maize. *See* maize
mutation and, 104, 106, 109, 159,
  241, 304
natural selection and, 96, 102-104,
  109. *See also* natural selection
peppered moth as illustration of,
  96-111
prerequisites for, 30
recombination and, 87, 104, 106
understanding of, 32, 332
  limitations to, 111, 314-315
unpredictability of, 112
evolutionary clock, 26-27
evolutionary unity, 63

Falkow, Stanley, 228
France, 152
fruit fly (*Drosophila melanogaster*),
  22-23, 39-40, 62, 303, 318
injection of foreign DNA into, 183
Morgan's work on, 39-40
mutation in, induced by X-rays
  (Muller), 40

galactosemia, 184, 249
Galton, Francis, on eugenics, 38-39
gene(s), 19. *See also* DNA
array of, in prokaryotes and
  eukaryotes, 269
biological mechanisms in:
  defense against mutagens, 241
  DNA excision-repair system,
    244-246
  repair of mutational damage,
    242-247
chemical nature of, 50-55
"dances" of
  chromosomal, 70-93, 112. *See also*
    chromosomes; meiosis; mitosis
  Darwinian (evolutionary), 94-95,
    96, 110-112. *See also* evolution;
    heredity
  molecular, 55-59
    protein synthesis, 55-64, 70, 78,
      88, 112. *See also* protein
    replication, 65-69, 70, 74, 78,
      112, 242, 244, 246
    mutation and, 104-105, 106, 109,
      247
environmental damage to, 238-264.

See also mutation
insensitivity to, human, 261-262,
  264
expression of, 53, 64, 157, 183,
  196-197, 277, 278
possible phenotypes in, 157
genethic principle on, 25, 344
Hardy-Weinberg equilibrium and,
  107-109
human understanding of, 32
  modern, 34-40, 41-47, 95-96,
  138-139, 338
linkage in, 110, 319-320
location of, in chromosomes, 39-40.
  See also chromosomes
manipulation of, 46-47, 114-140
mapping of, in human genome,
  316-339. *See also* human genome
  deciphering
  family pedigrees as evidence for,
    319, 324, 330, 334
  integration of, with DNA
    sequencing data, 334
  sex linkage and, 320
  linkages in, use of, 319
  recombinant technology for,
    325-327, 328-339
  somatic-cell hybridization for,
    323-325, 332
Mendel's discovery of, 35-38, 39, 45,
  95
molecular nature of, 32, 42
Morgan's work on, 39-40
Müller's work on mutation in, 40
origin of, 25-32
"primal gene," 25-30
  fundamental properties of, 30
  replication in, 28, 29, 30
recombinant, technology of. *See also*
  DNA, recombinant technolgy of
  cloning of, 116-132, 182, 318, 325
    cutting, with restriction enzymes,
      118-122
    growing, in cells, 126-128
    in plants, 275
    isolating desired gene, 128-131
    splicing, 118-124, 325
    transporting, to cells, 124-126,
      195-196
      crown gall bacterium as vector
        for, 275
      modification of vectors for,
        125
  gene therapy and, 182-185. *See also*
    gene therapy
  genetic screening and, 164-165
  military interest in, 217

absence of, on Asilomar
  Conference agenda, 229
biological weapons and, 211-212,
  216-217, 219, 220-235. *See also*
  biological weapons; military
potential usefulness of, 138-140,
  318
sickle-cell anemia and, 193-196
synthesis of, 136-138
transfer of, between species, 265,
  275, 276-279, 281, 285-286
  genethic principle on, 265, 345
recombination of, meiotic, 80-93,
  104, 106, 108. *See also* meiosis
regulatory, 64, 333
sequencing of. *See* DNA, sequencing
  of
variations in, future discovery of,
  329-330
gene insertion, 184, 194
gene modification, 184
gene surgery, 184, 314
gene therapy, 183-207
  candidates (diseases) for, 184,
    185-186, 193, 194
  germ-line, 181, 200-207
    feasibility of, 202-204
    future of, 206-207
    genethic principle on, 181
    perils of, 204-206
  sickle-cell anemia and, 186-196, 200
  germ-cell repair, 200, 202
  somatic-cell, risks of, 196-199
  strategies for, 184-185
genethics
  future of, 345-349
  issues of, 343-346
  principles of, specific, 25, 141, 160,
    181, 208, 238, 265, 290, 316,
    344-345, 347
genetic code, 31-32, 42, 44, 50, 60-64
  action of, in cells, 62
  ambiguities in, 62-63
  biological significance of, 63
  defined, 61
  evolutionary unity confirmed by, 63
  recombinant genes, use of, in finding,
    130-131
genetic "dances." *See* chromosomes;
  DNA, dynamics of; genes, dances of;
  protein
genetic diversity. *See* genetic variability
genetic drift, 106-107
genetic engineering, 114-132
  agricultural, 139, 267, 274-279, 303,
    312. *See also* crown gall bacterium;
    maize; monoculture

benefits of, potential, 139, 279
genetic uniformity and, 312-313
in maize, 290-291
moral questions in, 280
risks of, 280-281
technical problems with, 276-279
early attempts at, 33-35, 114-115
of germ cells, ethical dimensions of,
200-207. *See also* germ cells
military applications of, 212, 221-237.
See also biological weapons;
military
modern, 115-140
in higher animals, 182
promise of, 138-140, 182
moral principles of, need for, 236,
346-348
risks of, discussed at Asilomar,
227-229
in rodents, 182, 202
unpredictability of, 112
Wald on, 265
genetic exchange, 265, 345
bacteria, in, 282-283
between prokaryotes and eukaryotes,
268-269, 273, 274, 278, 288
barriers against, 265, 269, 278, 282,
284-285
rarity of, 287
between taxonomically distinct
species, 282, 334
barriers against, 284-286
recombinant techniques and, 313
cannibalization of raw DNA as
avenue for, 283
genethic principle on, 265, 345
maize, in, 296-297, 299, 307
genetic migration, 107
genetics
central dogma of, 53-55
defined, 19
emergence of, from Mendel's work,
39
future of, 45-47
history of (summary), 44-45
insights from, on being human, 341
mapping of human genome, as
challenge of, 317, 328, 334, 338
moral questions arising from, 180,
199-200, 208, 212, 235, 236, 280,
343-349
population, 94-95
genetic screening
adult, non-occupational, 168
defined, 162
germ-line gene therapy and, 205
human genome mapping and, 328-330

infant, 166-168
PKU and, 167-168
occupational, 162, 168-180
monitoring, for job-related genetic
damage, 169
perils of, 170-178
potential for good in, 174-176, 179
power imbalance in, 176-180
susceptibility to occupational
hazards and, 168-169
discrimination and, 169
prenatal, 152, 157, 166, 173
amniocentesis for, 166
emerging techniques for, 166
sickle-cell trait and, 162-163
strategies of, 163-165
Tay-Sachs disease and, 164
XYY genotype and, 152-154
genetic variability (diversity, variation),
30, 82, 95
biological weapons systems and, 218
Darwin's principle of, 96
evolution, as essence of, 83, 87, 95,
102-107, 205, 286, 300-301, 313
meiotic recombination as source of,
89, 104, 304
mutation as source of, 102-106,
205, 241, 250, 304
genethic principle on, 290, 345
genetic screening strategies and, 165,
173, 346
occupational, 170-173, 177
pathogens and, 219, 311, 313
plant gene pools and, 288
loss of, in domestic crops, 290,
300, 307, 312-313
hybridization and, 309
in maize, 300, 302, 304, 307
maize as example of, 304, 305-306,
314
preservation of, 313-314
value of, subjective, 303-304
population genetics and, 94-95
significance of, future, 329-330
Geneva Protocol, 213
Gerald, Park, and XYY screening, 153,
155
Germany
biological warfare and, allegations of,
213
Nazi, eugenics in, 41, 44-45
nerve gas and, 214
germ cells, 35, 48, 71, 80
discovery of, 35
division in (meiosis), 80-93. *See also*
meiosis
genetic variability in, 314

ionizing radiation and, 254, 256
manipulation of, 181-184, 198, 200-207
  ethical dimensions of, 200, 207
  feasibility of, 202-204
  genethic principle on, 181, 345
  promise of, 203-204
  resolution on, in *Congressional Record*, 181
  risks of, 198
  Romer on, 181
Mendel's work on, 35-38
repair of, 200, 202
  sickle-cell anemia and, 200
somatic cells and, boundary between, 203
stability of recombinant genes in, 278
"supermice" and, 203
toxins in food or water and, 262
Gilbert, Walter, on human genome sequencing, 316
Great Britain
  biological weapons research in, 214
  class divisions and hereditary differences in, Galton on, 39
  England
    industrialization in, 98-100, 101
      antipollution regulations, 101
      peppered moth in, 97-101
    genetic screening for XYY genotype in, 153
    karyotype surveys of mental patients in, 145-146
  Scotland, 145
  soldiers of, and biological warfare against North American Indians, 212-213
  soldiers of, as subjects of biological weapons research, 215
Greece, ancient
  beliefs about human heredity in, 33-35
    Aristotle's speculations, 34, 45
    Plato, writings of, 35
  biological warfare in, 212
  hubris in, concept of, 340
Green Revolution, 291, 300
Guthrie, Woody, 186

Hardy-Weinberg equilibrium, 107-109
  peppered moths and, 108-109
Heller, Joseph, 238
hemoglobin, 159, 189, 192
  importance of maintaining stable levels of, 198
  S, in sickle-cell disease, compared with normal, 189-191
hemophilia, 186

hereditary disease, 144, 162-174, 183-207, 330, 346. *See also specific diseases*
  "curing," complexities of, 202, 206
  eradication of, dream of , 200
  gene therapy and, 183-207
    candidates for, 185
    fatal, sufferers from, as candidates for, 199
  genetic screening for, 162-168, 170-174
  molecular, discovery of, 189
  monogenic, 184
  mutation as trigger of, 248-249
  new conceptions of, 206, 318, 329-330
  treatment of, without genetic intervention, 205, 346
heredity. *See also* evolution
  complexities in, 156-157, 314
    limits to understanding of, 337-339
  control of, early attempts at, 34-35
    artificial selection as, 35, 114-115
    Egypt, in, 35
    Plato on, 35
  Darwin's principle of, 96
  DNA and, 50. *See also* DNA
  early human understanding of, 32-35, 45
    agriculture and, 33-34, 45
    Aristotle's thoughts on, 34, 45
    Babylonian culture and, 33, 45
    cultural symbols for, 34
    Greece, ancient, and, 33, 35
    Hindu texts on, 35, 45
    India, early writings in, 34-45
  early life forms, in, 26, 31
  environment vs., 149, 151, 154, 157, 171, 179
    in genetic disease, 249
  eugenics and, 38
  evolution and, 30
  Hardy-Weinberg equilibrium and, 107-109
  human, genethic principle on, 316, 345
  imperfection, as characteristic of, 205, 250
  insights into, from human genome mapping, 330
  limits imposed on, by nature, 115
  Mendel's work on, 35-38, 45, 95
  modern understanding of. 19, 45-47, 95
  in populations, 94-95
Herrick, James B., 187-188
heterosis, 309
Hin Tijo, Joe, 143
Hiroshima, 23

Nagasaki and, 255-256
Hitler, Adolf, 40, 214
human beings (*Homo sapiens*), 26,
   32-34, 47, 80, 82, 96, 111, 301
  breeding experiments on,
    unacceptability of, 319
  complexity of systems in, 337-339
    rapid DNA sequencing, capability
     of, 327, 328
    risks of, 334-339
    sequencing of, 318, 343
    somatic-cell hybridization, for
     mapping of, 323-325, 332
     limitations of, 325
Human Genome Initiative, 44, 165, 328,
   329, 334
  multidisciplinary effort as alternative
   to, 335
human genome, deciphering, 44, 139,
   165, 316-340
  benefits of, 328-334
    to applied genetics, 328-330
    to basic research, 330-332
    to understanding evolution, 332-334
  geography of, 317-319
  recombinant DNA technologies and,
   325-327, 328-338
Huntington's chorea, 186, 330
Huxley, Aldous, 160, 171

immune system
  ADA deficiency, 186, 193
  AIDS, 232-235
  AIDS-like virus, use of as example,
   232-235
  PNP deficiency, 186, 193
  sickle-cell trait and malaria and,
   201-202
  targeted by biological weapons,
   potential for, 233
Inca civilization
  maize and, 293, 296, 297
India
  caste system in, 34
  early beliefs about human heredity in,
   34. 45
Indians. *See* North American Indians
  compulsion to understand in, 32, 342
  emergent properties in, 337
  environmental toxins and, 175,
   250-264
  gene library of, 131-132
  genetic privacy in, genethic principle
   on, 160, 345
  hereditary diseases or defects in,
   144-145, 162-168, 183-207
   gene therapy for, 183-207

genetic screening for, 162-174
  occupational, 168-174
insensitivity of, to genetic damage,
  261
maize and, bond between, 297-299,
  311
multidimensional nature of, Jacob
  on, 339
mutations in, monitoring for, 260-261
need for belief system in, 34, 341
place of, in global ecosystems, 342
use of, as subjects of biological
  weapons research, 215

industrialization
  environmental mutagens and, 241,
   250, 258, 262
  genetic risks of, 264
  peppered moth and, 98-101, 102, 106
industrial melanism, 98, 100-102, 106,
  111
industry, as major user of scientific
  knowledge, 21

Jacobs, François
  on logic of biology, 338-339
  on "objective" science, 339
Jacobs, Patricia, on XYY genotype and
  aggression, 145-151
Japan
  biological weapons research in,
   214-215
  industrial melanism in, 101
"jumping genes," 303
Jung, Carl, 340

Karyotyping, 75, 143-145, 150, 151
  detection of radiation-induced
   abnormalities by, 255-256
  Down's syndrome and, 144-145
  Klinefelter's syndrome and, 145
  Turner's syndrome and, 145
  XYY genotype and, 144-147
   controversy over, 150-151, 153-155.
   See also XYY genotype
Khorana, H. Gobind, 137
King, Jonathan, 154
Klinefelter's syndrome, 145
Krimsky, Sheldon, 229
Kundera, Milan, on wisdom, 316

Lancet, and XYY report, 142, 145
Laos, 216
Lapland, and Chernobyl accident, 254
Lesch-Nyhan syndrome, 166, 185-186,
  193
Levan, Alberta, 143

Maize (*Zea mays*), 272, 290-315
  artificial selection in, 295, 296, 301
  domestication of, 294, 295, 296, 300, 302
  evolutionary origins of, 293, 294, 295, 301, 302, 304, 305-306
  genetic uniformity in, 291, 302, 307, 310, 311, 312
  genetic variability of, 294, 301, 302, 304, 305
    destruction of, 300, 302, 307, 309, 311-312
  humans and, bond between, 297-299, 311
  hybridization of, 307-312
    genetic vulnerability and, 310-312
  mythology and religion of, 290, 297
malaria, and sickle-cell trait, 201-202
manic depression, 330
Maya civilization
  maize and, 293, 296, 297
  myths of, 347
McClintock, Barbara, 303
meiosis, 80-93
  defined, 71
  females, in, 83-91, 92
    polar body in, 90, 91
  gene linkage and, 319
  genetic variability and, 82-83, 89, 104, 106, 304
  maize, in, 304, 305-306, 314
  males, in, 91-92
  X and Y chromosomes in, 92
  mathematics of, 82
  nondisjunction during, in XYY genotype, 142-143, 169
  plants, in, 278
  recombination through, 86-88, 89, 90, 104, 106, 108
    Hardy-Weinberg equilibrium and, 108
  significance of, 82-83
  stages of, 84-93
Mendel, Gregor, 35-38, 39-40, 95. *See also* Mendelian genetics
Mendelian genetics, 35-38, 165
  dominance in, 36
  genes, discovery of, 36, 95
  Hardy-Weinberg equilibrium and, 109-110
  inheritance in, rules of, 37-38, 111
  separation of alleles in, 37
  "supergenes" and, 110
  underlying mechanisms, clues to, 38
Meselson, Matthew, on biological weapons, 230
Mexico, 295

Miescher, Johann Friedrich, 42
military
  biological weapons and, 209-331. *See also* Biological weapons
  scientific knowledge, as major user of, 21
Military Review, 223
mitosis, 71-80
  bacterial (binary fission), 267
  compared with meiosis, 84, 86, 89, 90
  defined, 71
  disruption of, by nuclear explosion, 255
  monitoring, for human germ-line therapy, 204
  mutation and, 249, 262
  red blood cells and, 193
  role of, in XYY genotype, 142
  significance of, 79
  stages of, 72-78, 79
  "supermice" and, 203
monoculture, 226, 291, 300, 307, 312, 314
monogenic disorders, 184
  examples of, 184, 185-186
    potential for gene therapy in, 184, 193
  responsibility of, for serious diseases, 249
  transmission of, triggered by mutations, 248-249
moral values
  cross-cultural synthesis of, 348
  destruction of, and biological weapons, 235
  genethic principles, in, 343-344
  germ-cell therapy, in, 180, 199-200
  lack of, and military use of genetic technologies, 212
  myth, place in, Jacob on, 340
  need for, in genetic engineering, 236, 346
  proscribing human experimentation, 261, 319
  questions of, in agricultural genetic engineering, 280
  separation of, from science, 286-287, 341
  unacceptability of biological weapons and, 208, 218, 345
  Western, 347
Morgan, T.H., 39
Muller, Hermann J., 40
multi-X genotype, 145-150
  Carstairs study and, 147
  extra Y chromosome and, 149-150
  Klinefelter's syndrome and, 145

mental deficiency and, 149
mutagens, environmental
 chemical, 104, 239, 256-261, 262, 264,
  345
  damage from, delayed appearance
   of, 260
  diversity of, 239, 256-257, 258
  injection of, into environment,
   256-257
  interactions between, 261
  modification of, by metabolic
   processes, 259, 260
  sources of, 258
  tests for, 258-261, 263
   Ames test, 258-259
   animal tests, 260
   monitoring of human populations
    as, 260-261
 genethic principle on, 238, 345
 ionizing radiation, 104, 239, 251-256,
  262, 264, 345
  action of, on DNA, 251-252
  increase in, through human
   activities, 252-256
  nuclear war and, 254-256
 rising levels of, 241, 250, 256-258,
  262
 solar (ultraviolet, UV) radiation, 104,
  239, 242-247, 262, 345
  carcinogenicity of, 261
  increase in, through ozone layer
   deterioration, 250
 "spontaneous" mutations from, 239
 vitamins and, 264
mutation
 biological weapons and,
  unpredictability of, 218
 chromosomal, 248, 249
 diseases triggered by, 248-249
 dominant and recessive, 198, 249
 environmental, threat of, 263
 ethical neutrality of, in nature, 250
 genetic "markers," as, 326, 329, 330
 genetic variability, source of, 103-105,
  106, 304
 human contribution to, 250-264
  nuclear technology as example of,
   251-256
 insertional, 198
 insights into, from mapping human
  genome, 330
 maize, in, 304, 305-306, 314
 natural selection, favored by, 158,
  241, 247, 248
 nature of, 238-240
 peppered moth and, 104-105
 random, as cause of hereditary

disease, 205
 repair of, by genetic systems, 242-248
  excision-repair, 244-246
  postreplication repair, 247
 sources of, 238, 239, 341
 "spontaneous," 239
 X-rays, generated by, 40, 45
  early genetic engineering, in, 115
  fruit flies, in, 40
mythology
 Chimera in, to explain recombinant
  DNA, 123-124
 Cohen on, 114
 Haida Indians and, 287, 347
 hereditary, to explain, 32, 34
 maize, of, 290, 297
 science and, 341-350
  non-Western, 348-349

National Institute of Mental Health
 (U.S.), and XYY syndrome, 152-153
natural selection
 biological barriers and, 284
 biological weapons and, 222
 cooperative, 274
 Darwin on, 95-96
 dependence of, on mutation, 205,
  241, 247
 environmental toxins and, 175, 262
 genetic "defects" and, 158, 205
 genetic variability and, 103-106, 205,
  286, 313
 maize, in, 296, 304, 306
 peppered moth as illustration of,
  97-103
 universality of, 313
Nature, and XYY controversy, 145-148
New Scientist, and XYY controversy,
 154
Newsweek, and XYY controversy, 152
Nixon, Richard, 210
North America, 202, 219, 225, 292
 monoculture in, 291, 310
North American Indians, 224
 early beliefs among, about human
  heredity, 34
 Haida, cultural myths among, 287
 maize and, 291-292, 293, 297
Novick, Richard, 228
nuclear technology
 DNA damage and, 251-256
 war using, possibility of, 254-256
nucleic acids, 25-31. See also DNA;
 RNA

oral contraceptive, 18, 23, 342
Origin of Species, The (Darwin), 95

ozone layer, 23, 240, 241
  air pollution and, 250

Palmiter, Richard, and gene transfer,
  182, 202
Pan, 33
Pauling, Linus, and sickle-cell anemia,
  188-189
Paz, Octavio, on memory, 25
peppered moth (*Biston betularia*),
  96-110
  dominance in, 98-101
  industrial melanism in, 98, 100-102,
    106, 111
Pisum sativum (garden pea), 36-38, 318
PKU (phenylketonuria), 167-168, 249
plasmid(s)
  characteristics of, 272, 284
  resistance to antibiotics in, 272
  value of, for military purposes, 222
    hazards of, 228-229
  Ti (tumor-inducing), in crown gall
    bacterium, 271, 272-273, 275, 277,
    288
  vectors for gene cloning, as, 124-126,
    127, 128
Plasmid Working Group, 228, 229
  on biohazards of plasmid exchange,
    228
  on recombinant biological weapons,
    229
Plato, 35
pneumonic plague, "promise" of as
  biological weapon, 209
PNP deficiency, 186, 193
polygenic traits, 109
  agricultural, 276-277
  diabetes as, 186
  genethic principle on, 141, 344
  majority of inherited traits as, 157,
    206
  mutation and, 249
prokaryotes, 265, 269, 274, 278, 282,
  284, 286, 287
  compared with eukaryotes, 269-270
protein
  amino acids in, 30-31, 52, 331
  abnormal, in sickle-cell disease, 189
  central dogma and, 53-55
  chromosomes, in, 49
  eukaryotic DNA, in, 269
  function of, in gene replication, 68
    mutation and, 105, 244
  production of, by genes, 41-42, 44,
    52-53
  roles of, in cells, 52

  synthesis of, 55-65, 70, 78, 331

transcription of DNA sequences
  and, 55-58, 121
translation of DNA sequences and,
  58-63
genetic code as key to, 60-63

Queen Victoria, 186

radiation, 239
  ionizing, 239, 251
    action of, on DNA, 251-252
      delayed effects of, 252, 253, 255,
        256
    human sources of, 252-254
      Chernobyl as, 254
      nuclear war and, 254-256
  solar (ultraviolet, UV), 239
    damage caused by, in DNA,
      242-247
    mutagen, as, 239, 241, 242, 261
    ozone layer and, 240, 241, 250
radioactive fallout, 23, 255
radioactive labeling
  of DNA copies of RNA molecules,
    130
  genetic screening and, 164, 166
  of nucleotide bases, for gene
    sequencing, 132-137
  of RNA messenger molecules, 129
  of synthetic genes, 130-131
recombinant DNA. *See* DNA,
  recombinant; genes, recombinant
recombination, meiotic. *See* meiosis
religion
  agricultural practices guided by, 34
  genethic principles and, 344
  heredity, as explanation of, 32, 33, 34
  maize, as focus of, 297, 299
  non-Western, 348
  science and, 341, 348
reproduction. *See also* heredity
  cellular. *See* meiosis; mitosis
  differential, 101
  early human understanding of, 32-35,
    45
  menstrual cycle and, 82
reproductive cells. *See* germ cells
restriction enzymes, 118-122, 126
  DNA sequencing, use in, 132-134
  foreign DNA, as defense system
    against, 283-284
  genetic screening, use in, 164, 166
  mapping human genome, use in, 325,
    326
  mutations, use for detecting, 249, 326
retinoblastoma, 249, 322
retrovirus(es), 54-55

action of, 268
cancer-causing agents, as, 199
ethnic weapon, use of in hypothetical scenario, 232-235
genetic exchange between species and, 282
monitoring of, potential for, 331
RNA
central dogma and, 53-55
compared with DNA, 53
defined, 26
nucleotide bases in, 53
recombinant genes, as probe for, 129-130
role in protein synthesis, 32, 53-55
transcription of DNA sequences, 55-58, 70
translation of DNA sequences, 58-63, 70
Romer, Alfred Sherwood, on germ cells, 181
Russell, Bertrand, on ethics, 340

schizophrenia, 330
science
"Big Science," 328, 336
curiosity in, 334, 338, 343
enthusiasm in, 335, 338
ethics in, need for code of, 236, 237, 340, 343, 346, 347, 348
genethic principle on, 347
impact of, on modern society, 18, 20, 341
Jacob on, 341
limits of knowledge in, 22-24, 337-339, 341-343
oral contraceptive as illustration of, 23, 342
militarization of, 237
myopia of, 342
power and profit in, 20, 21, 176-180, 212, 280, 336
reductionist nature of, 19, 337, 341-342
emergent properties in, recognition of, 337
secrecy as enemy of, 237
social implications of new discoveries in, 21, 23, 342
social responsibilities of, 236
spiritual and moral values, separation from, 286-287, 299, 343
vitalism and, 337
Science for the People, 154
Sendai virus, 323
sex cells. See germ cells

sickle-cell anemia, 162
discovery of, 187-188
gene therapy for, 186-196, 200
bone marrow cells in, manipulation of, 193, 194
genetic exchange in, 194
germ cells in, repair of, 200, 202
genetic basis of, 188-191
hemoglobin, defect in, as cause of, 189-191
genetic screening for, 164, 166
prenatal, new tests for, 166, 192
medical management of, 192
Pauling's work on, 188-189
sickle-cell trait, 162-163
genetic screening for, by U.S. Air Force Academy, 162-163
malaria and, 201-202
Singer, Isaac Bashevis, on knowledge, 339
Sinsheimer, Robert, on biological weapons, 208, 230
smallpox, as biological weapon, 209, 213
somatic cells, 48, 71
division in (mitosis), 71-80. See also mitosis
genetic manipulation of, 181, 197, 198, 199, 200, 207, 345
germ cells and, boundary between, 203
hybridization of, 323-325, 332, 333-334
ionizing radiation and, 255
Sontag, Susan, 316
South America, and maize, 293, 294
Southern leaf blight fungus, 310-312, 313
Soviet Union, 212
biological weapons and, 212, 233, 236
accusations against, post-World War II, 216
Biological Weapons Convention of 1972,
as signatory to, 229, 230
funding of research on, reported, 217
Chernobyl and, 254
Speck, Richard, 152
Sperry, Roger, 337
strontium 90, 253
"supercrops," 279
supergenes, 110
"supergerms," 221-222
"superlivestock," possibility of, 183
"supermice," 44, 183, 203

Sweden, 149

Tay-Sachs disease, 164
teosinte, 294
Third World
  agricultural weapons and, 227
  dysentery in, 217
  genetic variability of plants in, 314
  preventable disease in, 346
  vulnerability of, to genetic toxins, 262
Thomas, Lewis, on DNA, 25
Tripsacum, 294
Turner's syndrome, 145

United States, 233, 248, 328
  Air Force Academy of, and genetic
    screening, 162-163
  biological weapons and, 210, 212,
    214, 215, 233, 236
    accusations against, 216
    accusations by, against Germany,
      213
    Biological Weapons Convention of
      1972,
      as signatory to, 229, 230
    destruction of weapons in, 210
    funding for research on, by
      government and military, 216-217
    research on, during World War II,
      214
    soldiers of, as subjects of Japanese
      research, 215
    use of, against American Indians,
      213
  blacks in, with sickle-cell disease, 187
  crown gall bacterium and, 266
  eugenics movement in, 40-41, 45
  genetic screening in, for XYY
    genotype, 151-153
    at Boston Hospital for Women,
      153-155
      controversy over, 154-155
    conference on, 153
    of juvenile males, 152-153
    in maternity hospitals, 153
  Human Genome Initiative and, 328,
    335
  industrial melanism in, 101
  maize in, 292, 294, 307
  mutations and, 263
  XYY controversy in, 152

virus(es)
  action of, 268
  AIDS, 322, 323
  biological weapons, use in, 209, 211,

216, 224
  hazards of, discussed at Asilomar
    Conference, 229
  hypothetical scenario of, 231-235
cancer ignition by, 199
cauliflower mosaic, 279
central dogma, as exceptions to,
  54-55
gene cloning, use of, as vectors in,
  124-126, 199
infections from, 262
nucleotide base sequence in, mapping
  of, 338
recombinant, in "gene library"
  construction, 132
resistance to, in wild corn, 295
Sendai, 323
vitalism, 337
vitamins, 264
  A deficiency, and blindness, 346

Wald, George, on genetic engineering,
  265
Wallace, Alfred Russel, 95
Walzer, Stanley, and XYY screening
  program, 153, 154, 155
Watson, James, 42, 50, 116
wisdom, 316, 340, 345, 348
  non-Western sources of, 348-349
World War I, 213
World War II, 20, 214-215
Wright, Susan, on biological weapons,
  208, 330

X and Y chromosomes, 92, 141-150
  aggression and, 148-149, 156, 159
  color blindness and, 319-320
Xeroderma pigmentosum, 246
X-rays
  carcinogenicity of, 261
  genetic engineering, using, 115
  ionizing radiation, as source of, 251,
    252, 255
  mutation in fruit flies, agent of, 40,
    45
XYY genotype, 142-159
  aggressive behavior and, 147-148,
    149-150, 151
    Jacobs on, in *Nature*, 145-148, 151
    Lancet on, 144
    sensationalism about, 151-152
    similar studies of, 151
  cause of, 142, 145
  controversy over research on, 150-151
  environmental factors and, 151, 154,
    157

discovery of, 142-143
  karyotyping and, 143-144
genetic screening and, 150, 152, 153,
  154-155
  controversy over, 154-155
  "The XYY Syndrome: A
    Dangerous Myth" (Beckwith and
    King) and, 154
lessons from debate on, 156-159
mental deficiency, association with,
  147-148, 150, 155
"supermale" and, 149
powerlessness of subjects and, 151

yellow fever, 209
Yellow Rain, 216

Zea diploperennis, 295

CREDITS

The quotation on page 114 is from "The Manipulation of Genes" by Stanley N. Cohen (*Scientific American*, July 1975, p. 25) and is reprinted with the permission of the publisher.

The quotation on page 238 is from *Catch 22* by Joseph Heller (New York: Dell, 1961, p. 184) and is reprinted with the permission of the publisher.

The quotation on page 339 is from *The Possible and the Actual* by François Jacob (Seattle: University of Washington Press, 1982) and is reprinted with the permission of the publisher.

The quotation on page 265 is from "The Case against Genetic Engineering" by George Wald, and is reprinted by permission of *The Sciences,* September/October 1976 issue.